I0824364

PRAISE FOR *AFRICULTURE*

"In *Africulture*, fifth-generation Black farmer Michael Carter Jr. presents an uplifting and enlightening celebration of Black people and African crop varieties and agricultural practices that were, and are, critically important to the success of American agriculture. He educates using personal memories, cultural history, and stories of his success with African varieties like jute and Nigerian spinach.

"Throughout the book, Carter draws on the metaphors of a plant and a seed to discuss not only plant health and bounty, but also the larger issues for African Americans working to succeed in farming. *Africulture* points the way to using organic and sustainable practices, cooperative marketing, and community education to grow a new generation of increasingly more successful Black farmers and agriculturalists."

—**IRA WALLACE**, Southern Exposure Seed Exchange; author of *The Timber Press Guide to Vegetable Gardening in the Southeast*

"The ancestors are undoubtedly shaking their tambourines in celebration of *Africulture*, Brother Carter's reverent recounting of the noble, dignified, and expert contribution of Black people to American agriculture in the face of extinction-level threats. Carter elucidates the poetic kinship between the Black agrarian narrative and the botanical life cycle of the very plants we tend, interweaving his vulnerable personal memoir as a fifth-generation farmer. *Africulture* provides a blueprint for the blossoming of an agriculture rooted in cultural memory, ecological care, and mutual thriving."

—**LEAH PENNIMAN**, cofounder, Soul Fire Farm; author of *Farming While Black*

"*Africulture* speaks to Michael Carter Jr.'s unique perspective on food and farming, which is informed by: being a fifth-generation Virginia farmer; having lived and farmed in the United States and Africa; and being a student of history, culture, and spirituality.

Brother Carter challenges us to think about agriculture as a fundamental part of our work to rebuild the self-determination that was lost through colonialism and enslavement. Written to be accessible to those interested in and engaged in the work of building Black food sovereignty, this book incorporates humor, stories, and interviews that give insight into the world of Black food and farming. *Africulture* makes an important contribution to the growing body of literature about Black land and agriculture."

—**MALIK YAKINI**, cofounder,
National Black Food and Justice Alliance

"As a fellow Virginian, I am honored to offer praise for Michael Carter Jr.'s *Africulture.* Mr. Carter reminds those of us who need reminding that without African farming experience, white immigrants like me would not have survived long in the American colonies. He does this in a lively, engaging style, including metaphors of Black history as stages of plant growth. I appreciate his descriptions of tasty African crops that we could grow in Virginia for climate change resilience. With inspiring bios of prominent Black farmers and educators, *Africulture* provides an uplifting message and encouragement to other farmers, especially Black farmers."

—**PAM DAWLING**, author of *Sustainable Market Farming, Second Edition* and *The Year-Round Hoophouse*

AFRICULTURE

AFRICULTURE

HOW THE PRINCIPLES, PRACTICES, PLANTS, AND PEOPLE OF AFRICAN DESCENT HAVE SHAPED AMERICAN AGRICULTURE

MICHAEL CARTER JR.

FOREWORD BY MICHAEL W. TWITTY

Chelsea Green Publishing
White River Junction, Vermont
London, UK

First published in 2026 by Chelsea Green Publishing | PO Box 4529 | White River Junction, VT 05001 | West Wing, Somerset House, Strand | London, WC2R 1LA, UK | www.chelseagreen.com
A Division of Rizzoli International Publications, Inc. | 49 West 27th Street | New York, NY 10001 | www.rizzoliusa.com

Publisher: Charles Miers
Deputy Publisher: Matthew Derr
Project Manager: Yardena Carmi
Developmental Editor: Fern Marshall Bradley
Copy Editor: Diane Durrett
Proofreader: Ashley Davila
Indexer: WordCo
Designer: Abrah Griggs

ISBN 978-1-64502-301-2 (hardcover) | ISBN 978-1-64502-302-9 (ebook) | ISBN 978-1-64502-303-6 (audiobook)
Library of Congress Control Number: 2026004698 (print)

Our Commitment to Green Publishing

Chelsea Green sees publishing as a tool for cultural change and ecological stewardship. We strive to align our book manufacturing practices with our editorial mission and to reduce the impact of our business enterprise in the environment. We print our books using vegetable-based inks whenever possible. This book may cost slightly more because it was printed on paper that contains recycled fiber, and we hope you'll agree that it's worth it. *Africulture* was printed on paper supplied by Marquis that is made of recycled materials and other controlled sources.

Authorized EU representative for product safety and compliance
Mondadori Libri S.p.A. | www.mondadori.it
via Gian Battista Vico 42 | Milan, Italy 20123

Printed in Canada.
10 9 8 7 6 5 4 3 2 1 26 27 28 29 30

itYAHfaot

Dedicated to the spirit, energy, and memory
of my ancestors who are pleased with me, and to
Mekhael, Yahir, and Shmael as my greatest fruit and legacy.

CONTENTS

FOREWORD

If you didn't know it already, Africulture will convince you: American agriculture is impossible without Africa and without Black people. This conversation, well overdue, has searched for texts to represent not only facts but intention. The story of the relationship between humans and plants and the transfer of ethnobotanical knowledge from West and Central Africa in particular to the Atlantic world transformed the history of humanity and the fate of the world economy and world culture. And yet, it is not the story solely of a people enslaved and it does not end with abolition as Michael Carter Jr. makes clear. It is a narrative of family journeys, individual discoveries, and lives dedicated to translating experience, feeling, knowing, and spirit across generations.

In my own quest, I wrote *The Cooking Gene* almost ten years ago. Similar to Michael, it was important for me to know the plantations of the American South where my own family had forged this history with their own hands. I wanted to understand the story of land ownership and inheritance in my lineage; to go to West Africa to see for myself where we came from and to understand how the relationship with food and working with the land shaped the nexus of our culture. This is very personal work. The Ancestors are part of the earth and the water, and it pleases me to see this book include not only acknowledgment but libation. *Africulture* is a special book because it comes from a special culture.

Sub-Saharan Africa has over 2,000 wild food plants to its credit, thanks to millennia of experience. Millet, sorghum, African rice, okra, black-eyed peas, watermelons, muskmelons, coffee, kola, tamarind, sesame, and so many more foods have come to the global table even as foods from other parts of the world—Asia and the Americas—have morphed and moved across the African continent, emerging in greater diversity and flavor than they came in. Neither the clouds of colonialism nor enslavement can hold all of this narrative, because today the Black world is breaking free of those chains. In Africa and her diaspora, an entirely new generation has embraced this work to invigorate, taking on responsibility for safeguarding and celebrating their multi-faceted and rich agricultural heritage.

How bold for this book to come to life in the year celebrating 250 years of American independence, a break that would have been impossible without African and Indigenous agricultural experience. Maize / corn and other crops made local food ever present and accessible, while tobacco, rice, cotton, sugar, and indigo created sources for cash that would be the backbone of the American economy moving forward. Tobacco paid much of the debt to France for the American Revolution, cotton made the Industrial Age forge ahead in the North and South, and corn and wheat grown by enslaved hands made sure Europe did not starve during its revolutionary era. African Americans became inventors and innovators, and the earliest Black colleges emphasized a knowledge base that encouraged students to be a part of agronomy and the agricultural shift that was core to the philosophy of the New South.

Despite so many tragic and undermining elements offsetting our progress as a people, the future is as bright as our imagination can envision and behold. *Africulture* is a gift and inspiration. I hope that it will provide thought for food as we face new challenges and ask new questions. We stand on the shoulders of great Ancestors and have a great legacy to gift the generations to come. Our vision and the work of our hands matter; the soil and its fruits are our family.

MICHAEL W. TWITTY, author of the James Beard
Foundation Award–winning *The Cooking Gene*
January 2026

LIBATIONS

Libations are a tradition of pouring out a liquid to honor a deity, spirit, or ancestors. The liquid is rhythmically and slowly poured from a vessel as names are uttered with respect, reverence, and humility. I've witnessed this ritual performed at the beginning of many kinds of events, ceremonies, and meetings as a form of prayer and honor, humbly seeking permission from greater powers to proceed while formally recognizing their existence and importance. In the African American church culture in which I was raised, it was customary that before speaking in front of the congregation, you would give all honor and praise to the Creator, from whom all your blessings flowed. Cultures throughout Africa, Asia, North America, South America, and Europe use water, some form of alcohol, or other beverage to pay honor and respect to deities and ancestors. Libations called a drink offering were practiced among the priests in the Old Testament.

Below is my offering to many deities of cultures throughout the world as well as honoring and uplifting individuals in my life who I've known personally or THROUGH their works and deeds. I will never forget their impact on me and my life, and their names will be said, honored, and remembered in this book. Please take a moment to read these names and add the names of those people who have transitioned into the next realm, but whose lives, presence, works, or deeds have made an impact on your life.

Oya, Yah Yahwah, Ra, Nut, Nyame, Ogun, Anzar, Takhar, Tano, Apedemak, Olokun, Mukasa, Mawu, Oko, Shango, Aje, Obatala, Amadioha, Nana Buluku, Anyanwu, Modjadji, Denka, Asase Yaa, Yemaya, Mebege, Arebati, Roog, Qamata, Nhialic, Orunmila, Obaluaye, Ngai, Adroa, Anubis, Njoku Ji, Ikenga, Kibuka, Bumba, Aje Shaluga, Njambe, Kaang, Gu, Maher(u), Modimo, Age, Avrikiti, Amun-Ra, Bemba, Mbombo, Mukuru, Mulungu,

Mwari, Nzambi, Olorun, Olodumare, Osanobua, Ruhanga, Yah Shavuot, Allah, Elohim, Umvelinqangi, Unkulunkulu, Tabaldak, Gluskab, Tahquitz, Apsaalooke, Kokyangwuti, Muyingwa, Taiowa, Ha Wen Neyu, Whope, Kewkwaxa'we, Cautantowwit, Nana Tabor, Yotaanit, Asdzaa Nadleehe, To Neinilii, Nana Atrada, Tara, Drdha, Geb, Me, You.

We acknowledge the spirit of the legacy of those ancestors who worked with the Earth to sustain it, honor it, and explore and share her genius. We acknowledge those energies known and unknown that represent life, living in harmony with creation on this planet and beyond; those who masterminded and executed our liberation; and back to nature movements around the world.

A very personal appreciation to those spirits, elders, and ancestors who directly affected, influenced, and assisted in creating my life and the direction/trajectory thereof. I want to acknowledge everyone I can remember. If I have forgotten any, charge it to my head, not to my heart. These individuals drive, motivate, inspire, and influence me on a daily basis, and I pray you and spirits likened unto you continue to assist me in my goals and objectives and keep me aligned with my purpose.

Gran, Grandma Lizzie, Grandma Lucille, Grandma Rose, Uncle Keith Carter, Mr. Lewis Carter, Mr. Jefferson Davis Shirley, Mrs. Catherine Walker Shirley, Mrs. Sally Walker White, Mr. Robert Lee White, Grandpa Warren Carter Sr., Mr. James G. White, Grandpa Gil (Gilford White), Grandpa Eddie West, Grandpa Clifford, Grandma Mabel Fleming, Aunt Bernice Byrd, Aunt Marjorie Lightfoot, Aunt Virginia Carter, Uncle Gilford Bayboy White, Uncle Robert White, Uncle Isaiah White, Uncle Joe Carter, Uncle Roy Carter, Uncle John Scott, Uncle Lee Scott, Aunt Anna Scott, Nana Louise French, Uncle Al Burrows, Cousin Sean Jones, Cousin Darryle Pretty, Aunt Faye Pretty, Aunt Bettye Jones, Uncle J. R., Cousin David Carter, Aunt Ruth Carter, Aunt Evelyn Carter, Cousin Robert "Jeff" Barbour, Cousin Jonah West Ramirez, Cousin Josh Ramsey, Ms. Jada White, Cousin Michael Fleming, Uncle Eddie "Junie" West, Aunt Gloria Lovelace, Dr. Edna Lewis, Deacon Floyd Bates, Deacon Woodrow Bates, Deacon A. B. Howard, Deaconess Howard, Uncle Carlton Lewis, Mr. Rex Harris, Aunt Carrabel Garnett, Uncle Eddie Garnett, Aunt Sallie Johnson, Uncle Robert Keeves, Mr. John Wright, Mr. Ray Wright, Mr. Lomax,

Mr. Spencer Gwaltney, Uncle Little John Johnson, Dr. Gibson, Mrs. Gibson, Mrs. Coleman, Mr. and Mrs. Roane, Aunt Mary Ann Lewis, Mr. Harold Byrd, H. E. Ben Ammi Hamasheak, Sgan Ahtur Dahveed, Nasik Shaleahk, Nasik Eleazer, Ahk Tsavaliel Malik, Ahk Ahr, Dode Beniel, Dodah Ahcote Shemriyah, Ahcote Zimriyah, Dode Ahk Azriel, Ahcote Toomah, Ahk Tsavaliel, Ahk Yehudah, Dode Ahk Yehoyadah, Ahk Ahvdiel, Nasik Daniyel, Ahk Ben Hosheah, Ahk Yoahk, Mr. Judge Graham, Mr. Michael Collins, Ahcote Yaheerah, Mr. Ian Davis, Mr. Brandon Tolliver, Mr. Brian Scott, Mr. Tyrone Johnson Jr., Ms. Patricia Smith Wade, Ms. Florence Smith, Hon. Marcus Mosiah Garvey, Hon. Nat Turner, Bishop McNeal Turner, Hon. Booker T. Washington, Dr. George Washington Carver, Dr. W. E. B. Dubois, Hon. Patrice Lumumba, Hon. Tetteh Quarshie, Pres. Kwame Nkrumah, El Hajj Malik El Shabazz, Hon. Medgar Evans, Baba Dick Gregory, Dr. James Farmer, Baba Tarik Oduno, Dr. Huey P. Newton, Hon. Fred Hampton, Hon. Kwame Ture, Hon. Sekou Toure, Hon. Dr. Namdi Nzekwi, Hon. Khalid Muhammed, Dr. Amos Wilson, Dr. Ben Yochanan, Dr. John Henry Clarke, Hon. Paul Robeson, Hon. Bob Marley, Hon. Dr. Major Martin Delaney, Hon. Dr. George Padmore, Dr. John Hope Franklin, Hon. Walter Rodney, Dr. Jawana Kunjufu, Hon. Nannie Helen Boroughs, Mr. Miles Davis III, Mr. John Coltrane, Mr. Cannonball Adderly, Mr. Michael Jackson, Hon. Harriet Tubman, Hon. Mayor Marion Barry, Mr. Prince Rogers Nelson, Mr. William Withers Jr., Mr. Isaac Hayes, Mr. Slyvester Stone, Mr. John "Redd Foxx" Sanford, Mr. Richard Pryor, Mr. Fela Kuti, Mr. Marvin Gaye, Mr. Donnie Hathaway, Mr. Joe Sample, Mr. Michael D'Angelo Archer, Sar Shlomo, and countless others who shaped and molded my thoughts, experiences, family, and life.

It's noted in various cultures that a person doesn't truly die until you stop speaking, writing, or remembering their names. Let these individuals, and so many more righteous and courageous others live on in legacy to support our efforts on this side of creation. *Asé Asé Asé ooo. Melah.*

INTRODUCTION

If you find a book you really want to read,
but it hasn't been written yet, then you must write it.
Toni Morrison

Sometime in the 1930s, my great-grandmother Mrs. Mattie V. Carter stood in a county office building in Orange County, Virginia, and made a commitment to save and prepare a legacy for her family.

That legacy was land already in the family. It was a 150-acre farm that Mrs. Carter's parents, Mr. Jefferson Davis Shirley and Mrs. Catherine Walker Shirley, had purchased back in 1910.

The family, like most in rural Black Virginia, had farmed vegetables, cows, chickens, and pigs, and the property had provided security for their large family in the American apartheid era. It was a place of rolling fields and shady woods, deep in a rural county where small hamlets and African American–styled villages are scattered among farms. The fourth US president, James Madison, had lived nearby, on a plantation worked by the dozens of people he enslaved, who possibly included some of my ancestors.

The land where the young Mattie had grown up became the subject of a court case during the Depression; the family owed money on the property and most of my great-great-aunts and uncles didn't want to pay off the debt. But Mrs. Mattie Virginia Carter committed to securing the land. It wasn't an easy promise to fulfill, taking on the financial responsibility of a 150-acre farm and molding a future for the next four generations—but building or adding to legacies never is.

My great-grandmother (who we lovingly called Gran), a mother of twelve, had a grace and fluidness about her that I admired as a young person. She never seemed to stress, smiled caringly with her large round eyes (that most

At the family farm, 1959.

of us Carters still possess) and always opened her home for family business, counsel, and visiting. The latest issues of *Jet* magazine were there on her coffee table, and I always made it my responsibility to be read up on the centerfold. I spent a decent amount of time with Gran, watching her fry chicken and gizzards, stew cabbage, bake cornbread, and make sweet tea, and listening to her and my great-aunts chat about the grown folks' business. I remember spending hours in the front yard, playing what seemed to be a never-ending game of fetch with my first canine friend, Lady, her spaniel mix.

Gran's two-story white house was the Carter family headquarters. A majority of the family members made a yearly pilgrimage to Gran's, trekking slowly down the half-mile road on Carters Lane to her home. I'm quite sure my first introduction to the family came on that porch, being greeted by someone rocking on the metal rocker. By securing the land, Gran also secured the future of the family and a sliver of legacy that I share in this writing.

Portions of my ancestry can be traced back to the first generation of exiled Europeans arriving on the shores of what they described as the New World.

The documentation of most of the generations I've been able to trace identifies them as farmers or laborers. Some of my ancestors may have been from Angola, specifically the Mbundu nation and possibly a subsect of the Akans from the region with the imaginary boundaries derived during the Berlin Conference in 1885, which created the nations we know today as Ghana and Ivory Coast. They were brought to Virginia in the 1600s or 1700s and enslaved on plantations in the newly established Virginia colony. Some potentially were enslaved at the Shirley Plantation, the oldest plantation in America. And although I have no documentation for it, I have been told I have Indigenous American heritage that would precede my African and European ancestry in this nation. I've been told since I was a toddler that my tall, straight-haired, copper-skinned maternal great-grandfather, Clifford Fleming, was "Indian" (the language that was used in the 1980s). My great-grandmother Grandma Lizzie, having copper skin tone as well, also has Indigenous roots—and her mother, seen in the only picture I have of her—exemplifies her Indigenous features.

I don't know in what year many of my ancestors arrived here because enslavers didn't keep detailed personal records of the individuals they enslaved. They were prisoners of war, whose crimes were a tolerant resistance to the heat of the sun, a genius for working with nature to grow crops, and a cursed resilience that forced them into enduring horrific discrimination, indifference, and vitriol, generation after generation. Their true names, ages, lineages—none of it was recorded with accuracy. They were treated as cattle. Actually, they were treated worse—cattle breeds such as Angus, Belted Galloway, and Jersey are identifiable by their physical characteristics; their homes and origins are known. Are we nearly as familiar with the Mende, Mbundu, Igbo, Temne, Mandinka, Fulani, Bakongo, Fon, Wolof, Chamba, and countless other African nations and states (not tribes)? Their citizenry and legacies experienced generations of terror (much like the Indigenous nations of the Americas), never before and hopefully never again seen and experienced in world and universal history.

The history of Black farmers in America is a story of oppression and resistance, violence and solidarity, sustenance and stewardship, glued and mortared together with historical self-worth, family pride and cohesion, African principles of right and sensibility, and a belief in the greatness of our ancestors and the Creators. Using over 4,000 years of agricultural knowledge and innovation, we've made untold contributions to the development of American agriculture, not just through labor but through innovation and traditional knowledge.

My great-grandfather Clifford Fleming.

But after all those generations, Black farmers are on a steep decline. As I write and edit these passages, new threats to Black farmers have been legislated, policy-adjusted, or executive-ordered into place. At the turn of the twentieth century, when Gran was just a toddler, Virginia had more Black farmers than any state in America, at over 25,000. When Gran was a young mother, starting her family in 1925, there were a little more than 50,000 Black farmers in Virginia, again leading the country. By 2017, when I officially founded Carter Farms here on the land she saved, there were only about 1,300. That's a 98 percent decline—an extinction-level event. And in any situation where you have extinction, the cause is likely the loss of favorable conditions in an environment. To save the affected species, you must change the environment.

Carter Farms and Africulture, a nonprofit organization I began in 2020, seek to grow farmers of African descent as part of a larger effort to fight off an extinction-level event of civilization's foundational profession, indeed the world's oldest profession. We're incubating new farmers, helping older farmers grow their markets, and spreading the message that US agriculture could not have existed without African principles, practices, plants, and people. This book shares various changes and realities and conditions that have affected the existence and growth of Black farmers in America. That list includes a few other p's that are just as prominent, if not more so: policy, psychology, and perception: policies that have challenged the growth and success of many Black farmers, and psychologies, or states of mind and perception cloaked in prejudice and stereotypes. These challenges have shaped views around agriculture. Many residues of those policies and psychological perceptions still exist today.

At the time I was growing up in the 1980s and '90s, older farmers would tell me: "I don't make no money farming." If you want to discourage a teenager from entering your vocation, tell them you don't make any money doing it.

Table 0.1. 1900 Negroes in the United States Census

State	Number of Negro Farm Owners
Virginia	26,566
Mississippi	21,973
Texas	20,139
South Carolina	18,970
North Carolina	17,520
Alabama	14,110
Arkansas	11,941
Georgia	11,375
Tennessee	9,426
Louisiana	9,378
Florida	6,552
Kentucky	5,402
Maryland	2,262
West Virginia	534
Delaware	332
District of Columbia	5

Prepared by W. E. B. Du Bois. Appears in a bulletin from the United States Bureau of the Census, *Negroes in the United States*, published in 1904.

I was in 4-H and FFA (founded at Virginia Tech in 1925 and formerly called Future Farmers of America), but I never found them attractive because I never saw Black people participate in those activities. I'm not saying there aren't Black members in the organization, because there are, I just never encountered them at any of the events I participated in. And even though I majored in agricultural economics in college, I didn't plant one seed for the entire four years. Like many agriculture students I know who attended HBCUs, I was taught and encouraged to work at John Deere, the USDA, or another agribusiness, not how to be a farmer.

These are just some of the reasons why there's been a loss of 98 percent of Black farmers in my state and throughout the country since my great-grandmother's time. But there's another reason, too. The contributions of

Table 0.2. Virginia Farm Census

Category	Census of 1945	Census of 1940	Census of 1935	Census of 1930	Census of 1925	Census of 1920
Full owners	124,363	113,510	121,490	104,956	130,117	121,454
Part owners	12,219	13,164	16,649	16,148	13,470	14,909
All white operators	138,096	139, 795	154,421	130,937	143,576	138,456
Full owners	104,523	94,580	98,173	85,756	101,602	95,934
Part owners	8,970	9,844	12,304	10,900	8,842	9,480
All nonwhite operators	34,955	35,090	43,211	39,673	50, 147	47, 786

Adapted from the United States Bureau of the Census, *Census of Agriculture: 1945*.

Black people have been ignored and forgotten. We haven't received proper recognition as the true seeds of American agriculture, a recognition that could challenge negative perceptions. Very few people, including young Black people, are aware of the long history of African contributions and innovations in agriculture. The general perception of Black people on farms centers on slavery, sharecropping, and poverty—rather than self-sufficiency, resistance, individual and communal wealth, political power, or self-pride or self-worth. Agriculture and farming should be thought of as an opportunity to build the future—the future of what crops we grow, who grows those crops, how we take care of and maintain our bodies and nature, and how we'll find resilience in the face of climatic and environmental changes both geographically and socially.

Africulture is a term that pulls together many subjects that may seem unrelated. It refers to the history of growing crops in Africa—which goes back to the ancient Egyptians and beyond—and to the history of agrarian masters in Africa being kidnapped and trafficked. With a 100 percent tax on their labor, their progeny, and their history, they were shipped like cargo to create a commercial farming industry in a country later to be called the United States of America. It refers to the crops that are familiar in Africa today as well as the crops that helped make the United States a powerful nation rooted in agriculture with the critical and essential assistance of African and Indigenous innovators. It refers to principles and practices for agriculture, which can be used on a farm in Virginia today just as they've been used in Africa for centuries. If you wear jeans, drink coffee or tea, eat chocolate or vanilla, or enjoy an ice-cold soda, Africulture is *your* history. This is Hen Asem, which translates to "Our Story" in the Fante, a dialect of the Twi language of Ghana.

From sweet potato vines to tall amaranth and Nigerian spinach stalks, all of which are grown at Carter Farms, observing the life cycles of our crops lets us ponder what forces govern human health and allow human communities to thrive. One of the core Africulture principles is the Hermetic Law of Correspondence, which states "as above, so below": just as plants need healthy soil to grow, people need to be surrounded by health and support in their families, communities, villages, and society. During all the stages from seed to flower to fruit, the condition of the surroundings determines the quality and abundance of the crop.

This techno-crazed civilization, which a majority of humanity presently lives in, started with an AI revolution. But before there was artificial intelligence, or Allen Iverson and his killer crossover, there was *agricultural intelligence* and *African ingenuity*. All of us are products of this agricultural intelligence and African ingenuity that shapes our genes, our foodways, our relationships, our culture, our health, and even our spirituality.

Throughout various spiritual treks in Israel and America, and especially while living in Ghana, I learned that everything is connected. Anansesem are the stories of Ananse, the original Spider-Man, a major part of Ghanaian mythology and folklore. Spiders weave webs, *ntintan* in the Twi language of Ghana. Just like the World Wide Web and the fungal strands that connect a forest ecosystem, these connections both above and below influence all other beings between the firmament and the core of the Earth.

After growing up in Virginia, I made a critical decision to move to Ghana, where I lived for five years in the 2010s. After those five years, in 2017, the land that Gran saved spoke to me and convinced me to return to Virginia and reconnect with my family legacy. Some of what the land whispered ever so loudly in my ear and soul is shared in this book. I trust you will see, and most importantly feel, the ntintan I weave through this tapestry of metaphors, analogies, history, culture, produce, vegetables, agronomy, movies, music, mythology, policy, spirituality, and philosophy.

CHAPTER 1

SEEDS

PAST, PRESENT, AND FUTURE IN A CAPSULE

This is my story, this is my song . . .

Fanny Crosby, "Blessed Assurance," a hymn sung every other Sunday by a majority of Black Baptist church choirs.

And what did you not do to bury me,
yet you forgot that I was a seed?
Dinos Christianopoulos,
from *The Body and the Wormwood (1960–1993)*

Seeds are divinely designed to be the perfect metaphor. I frequently heard the religious saying "faith like a mustard seed" Sunday mornings growing up. A little bit of faith is all that is needed to begin your journey. Like a seed, as faith is watered, it will start to grow.

As the reproductive units of flowering plants, seeds are tiny time capsules, containing genetic, cultural, and social remnants of the past and propelling them into the future as they grow to their potential. In the African-based "conscious" community, specifically the Five Percent Nation (also called the Nation of Gods and Earths, an offshoot of the Nation of Islam founded by Clarence 13X) and the Rastafarians, adherents refer to their children as seeds. A man's sperm or ejaculatory fluid is often referred to as seed, and this term and context is also used in the Bible. My point is that the seed of plants and the seed of man have been synonymous since the time of the ancients.

The story of the seed parallels the story of African and Indigenous peoples in America. The seed goes through a great transformation, but no one ever asks the seed how it feels. Does it know that it's on a one-way trip to having its fruits and future seeds potentially consumed by others? Some of whom may not know, understand, or align with its particular purpose, its existence, or the right of its future seeds to exist, and will label its full-grown self as a weed, an undesirable plant.

Seeds are beginnings. Seeds have the responsibility and unusual reality of having to fulfill their purpose of growing into their mature form. No plant (or person) looks like the seed from which they grew. Their metamorphic growth is distinguished by features that are not indicated in the miniscule seed of their inception.

A dandelion seed, if it lands in any area where there's some semblance of water, soil, minerals, and sun, will most likely germinate and begin its growing pattern. I'm sure we have all seen plants growing in places where they're not wanted: in the seams or cracks of sidewalks, in gutters, on the bed of old pickup trucks that haven't moved for a season, or in a heavily herbicide-sprayed lawn.

The growing process, I assume, is painful, unexpected, and quite debilitating to the seed. The seed's lovely intact shell casing cracks open once it's surrounded by the combination of moisture and warmth. In a ravenous and violent way, it starts to grow a tail, what is called a taproot. This taproot is its initial means of eating and feeding itself. Its very life is dependent on the taproot and the other roots that subsequently grow out of it. Fungal hyphae present in the soil attach to the root and the root releases exudates—a combination of sugars, proteins, and carbohydrates that act as a food source for microorganisms in the soil. This food source starts the seed's first bartering system, trading its exudates with the local bacteria, fungi, protozoa, and billions of other microorganisms in its rhizosphere (a fancy word for the immediate surroundings of roots).

My father's family held their first large reunion in 1991, and they brought in a storyteller, an African griot wearing clothes from the Yoruba or Igbo peoples of Nigeria. He carried a jolly, inviting smile, like a train conductor about to take us on a never-ending journey through the history of ourselves. I was twelve or thirteen at the time and literally sat on the edge of my seat, watching and listening. The griot, which I later learned about in-depth while residing in West Africa, is more of a linguist and historian than a storyteller. The griot is a carrier and protector of culture, and sacred seeds. Storytelling is the tool of choice that envelops listeners with history, wrapped in moral and social lessons of the past. With a slight gasp before he started to speak, this great linguist bellowed, "This is a story that happened a loooooong time ago . . ."

Growing up, my family history usually was passed down informally. At our family reunions, I'd be a fly on the wall as the family elders discussed the family news of the day and snippets of our history. My mother's family and my father's father's family held their gatherings on the third Sunday in August; my father's mother's side held homecoming on the first Sunday in September. If I could go back in time, I would have had my father videotape all those sessions. At family gatherings on my mother's side, there was an eerie feeling in my grandmother's dimly lit kitchen, like a séance. I listened with a thrill, goosebumps coating my thin arms, when they started to talk about deceased family members.

Sometimes you have to piece together your family history, your roots, from little fragments. A Bible here, a census record there, a comment in passing. There's not always a lot of information. I've asked my Uncle John, Aunt Mamie, and Aunt Bettye about my great-grandfather John Lewis Carter, and their answers were just, "He was a good man; he didn't take any mess." The end. Learning more has been a long journey for me, and although I couldn't go back as an adult to those family reunions of the '80s and '90s, I've asked a lot of questions and found myself piecing together stories and histories like a focused detective.

Seeds represent a promise that time will pass and the right moment for growth will come along, sooner or later. They can lay dormant for days or decades and germinate when all the conditions are right. Over the years I attended other kinds of gatherings as well, seed-planting fields if you will, and listened to magnificent orators: His Excellency Ben Ammi Ben-Israel; Kohain Nathanyah Halevi in Ghana, who inspired my speaking style; Dr. James Farmer Jr., the leader of the Congress of Racial Equality (CORE) and a civil rights pioneer (and the subject of the movie *The Great Debaters*); Dr. Jawanza Kunjufu; Dr. Na'im Akbar; Dr. Edwin J. Nichols; and Dr. Claud Anderson, among others. They planted seeds in me: the importance of family, of communal understandings, and their relationship to history and experiences. An ideal I learned in Ghana, and throughout my travels, is that information and knowledge is not in silos. All things are interconnected like the web (ntintan) of Ananse the spider. Since 2017, after returning from what was supposed to be my forever sojourn in Ghana, I have been weaving my own ntintan, as well as being caught in an ancestral ntintan.

THE SEEDS OF AMERICA

We are all products of our environment. Our bodies are made of the food we eat. Our genes are made up of what our ancestors ate, and our lineage should reflect the vision and sacrifices of our ancestors as well as their foods. For African Americans, there's a connection between the cells in our bodies and the foods our ancestors ate when they were living in Africa. And for America as a country, there's a deep tie between the nation's history and economy and some of those same African crops. Shepherded by enslaved Africans and grown on American soil by African expertise, the seeds of plants like rye, millet, okra, and rice helped sustain new arrivals during the early years of colonization. And other African crops—like indigo and cotton—became the engine of a powerful new economy.

As scholars including Dr. Lerone Bennett Jr. and Dr. Ivan Van Sertima have detailed, Africans had a presence in North America well before Columbus. In the mid-1320s, the Mansa of Mali (*Mansa* is the Mandinka word for emperor or king), Mansa Abubukar II, had voyaged to the West, sending over 200 ships, along with himself. Many of the early kidnappers and/or exploiters (more commonly referred to as "explorers and conquistadors") journaled about meeting "Ethiopians" or "copper-skinned" people in the New World. There were also nations of people in the New World who had darker melanated skin tones but who had never been to or had interactions with people of African descent.

The connection between African plants and the American colonies, however, began with the movement of enslaved people across the Atlantic. For a three-month journey on a slave ship carrying 200 individuals who needed to eat, a lot of provisions were required (even though the kidnappers always expected a high level of loss). Trinkets like mirrors, buttons, cutlery, and clothes were used as currency to purchase captured human contraband along the West African coast. The traders of the enslaved would also buy or trade for provisions with these trinkets. The food that was available from the coast of Africa was primarily grains, starches, and beans, and such nonperishable items were ideal for an extended sea journey.

The coastline that stretches along the countries of Guinea, Liberia, Sierra Leone, and part of the Ivory Coast was referred to as the Grain Coast, named after the peppery spice called grains of paradise. Along that coast, traders could purchase starchy foods that would hold well in transit across the ocean: millet and rice. Okra is another food whose pods and seeds are very durable and plentiful. These were some of the foods that sustained people nutritionally during the "Beginning Passage" and the Middle Passage. (Many of us have heard of the Middle Passage, which the kidnapped and enslaved had to endure, but there was a beginning or start of the experience, which I've termed the "Beginning Passage," that was just as cruel, gruesome, and fatal as the Middle Passage.)

The people who had been stolen from their homes and enslaved managed to bring along seeds, too. There are stories of the enslaved braiding seeds into each other's hair for safekeeping during the journey. Maybe more important than the physical seeds was the knowledge the people protected—generational wisdom about these foods, how to grow them, how to cook them, and the properties and nutrition they held.

The first recorded account of enslaved Africans in the areas of Turtle Island (a.k.a. North America) that Europeans from England had occupied is from 1619. This was just a few years after the founding of Jamestown and the "starving time" of 1609–1610, when three-quarters of the Jamestown colonists died of malnutrition and disease, forced to eat rats, snakes, and even each other. The British colonial experiment came close to total failure because the English refused to humble themselves to Indigenous nations who knew how to sustain themselves in this environment. Historian and author Howard Zinn details in *A People's History of the United States*:

> Everything in the experience of the first white settlers acted as a pressure for the enslavement of blacks. The Virginians of 1619 were desperate for labor, to grow enough food to stay alive.

> Among them were survivors from the winter of 1609–1610, the "starving time," when, crazed for want of food, they roamed the woods for nut and berries, dug up graves to eat the corpses, and died in batches until five hundred colonists were reduced to sixty.[1]

It isn't plausible that the American colonies could have sustained themselves nutritionally and financially without exploiting this new skilled labor force of captured African indentured servants, as well as Indigenous people captured and forced to work against their will. Along with the enslaved Africans came the leftover grains their kidnappers had stocked to feed people during ocean transit, and some of those seeds became the origins of crops that exponentially increased nutrition for all these new arrivals. European crops such as broccoli, kale, and spinach couldn't survive through the sweltering dog days of summer in Georgia or the Carolinas. But African plants could survive just fine—rye, millet, rice, and jute (which produces not only fibers for twine, but also edible leafy greens). They became Americanized staples, the foundation for the growth of the colonies. On my Virginia farm, where we grow numerous varieties of African vegetables, visitors ask me frequently, "How do you grow these plants here?" The answer is, just plant the seed. During the summer, especially in the southern areas of the United States, the climate is not much different, in terms of soil temperatures and ambient temperature, than in West Africa. And the hotter the climate in North America becomes in the future, the more sensible it will be to grow these crops.

Can you picture a world without rice and blue jeans? In South Carolina in the eighteenth-century, rice, cotton, and indigo were grown on enslavement camps called plantations. These crops were produced by the labor of enslaved Africans and Indigenous peoples who held generations of knowledge about how to cultivate these plants. Rice provided a nutritional staple, grown and prepared by enslaved labor, while indigo was an extremely valuable cash crop, more profitable in that century than sugar or cotton. Crops like these galvanized the colonization process. They were the key to the colonies' ability to build a financial engine powerful enough to subdue the Indigenous nations and, eventually, to separate from the colonizing country. And the crops thrived thanks to the highly skilled workforce, which carried master's- or doctoral-level knowledge (if such degrees had existed at the time), that was brought to the colonies to innovate agricultural production.

These African seeds, tended by African people on American soil, were the seeds of civilization as well as the seeds of destruction and the seeds of pain. From them germinated generations of people and practices, all the power and the sorrow of American history. Without African crops, expertise, and innovation, American agriculture and America itself would not exist.

THE RHIZOSPHERE

When I was younger, I did not intend to be a farmer. I had other plans—I was going to be a biotech engineer and invent a kind of lawn grass that would grow only a half-inch tall. Every child would thank me for eliminating the occupation of grass cutting. Later I thought about being an investment banker and a fashion designer. But when you're young you don't realize all the seeds that are being planted in you. Unknowingly, I had eleven or more generations of agriculturist genetic material coursing through my arteries. My father and his brother both majored in agriculture at Virginia State University. My great-uncle Mr. Buford White had paved the way a few years earlier, majoring in agricultural education and becoming the first agriculture teacher in the family. My father followed in those size-thirteen footsteps, which made me an agriculture student almost from birth. I was inundated with it: Many of the events I went to during my childhood were ag-related. My godfather, Mr. James Gibson, was an agriculture teacher, and his father, Dr. David Gibson, was an agricultural professor at Virginia State University.

I majored in agricultural economics at North Carolina A&T State University (AGGIE PRIDE!)—a marriage between my mother's degree in economics from Mary Washington College and my father's agriculture interests. But I also had a minor in fashion design, and after I graduated I went to work for a Washington, DC, fashion company started by my cousins Aaron Lydell Byrd and Nkrumah Kenyatta Byrd, called Indefinite Designs, home of Dalinkwent Wear.

In my later days of working with Indefinite Designs, I ended up researching how to grow cotton to make our own T-shirts. My research landed me with the United States African Development Consortium (USADC). Its creator-founder-CEO, Dr. Brimmy Olaghere, a Nigerian econometrician, planted more seeds in me—a devoted appreciation for Africa and our relationship with her. He shared with me that in the early 1960s he had been an aide for Ghana's first president, Osagyefo Dr. Kwame Nkrumah;

Dr. W. E. B. Du Bois; and Nigeria's first president, President Nnamdi Azikiwe, at the dawn of the African independence movement. I'd sit at Dr. Olaghere's feet, mesmerized with his recollections of the history of pre-independence and post-independence Africa.

That might not have had much to do with agriculture, but it formed the impetus for how I view family, community, and business success. Indefinite Designs, and the leadership of my cousin Lydell, planted a seed; they provided opportunities to young Black people and encouraged entrepreneurship. He and others like him weren't scared of providing opportunities or showing support to these young people. These young men and women, including myself, weren't seen as weeds, but seeds that needed to be placed in rich soil. The ntintan web of family, of principles, of values and support for your own people, planted in me the seeds I presently use to support a modest group of Black farmers and future farmers in the mid-Atlantic region of the United States.

Washington, DC, in the 1990s was full of givers of knowledge and information. I sat at the feet of so many elders, heard so many lectures, and shared meals with some intellectual and cultural giants. All of these experiences and many more were part of my rhizosphere. For a plant, the rhizosphere is the narrow region of soil or substrate that is directly influenced by the plant's root secretions and soil microorganisms (a.k.a. the root microbiome). The rhizosphere forms after the roots are established, but those microorganisms also are present in the soil before the seed germinates—doing their thing before the seed even shows up. And not to come empty handed, the seeds carry their own microcolony of microorganisms, too. The microorganisms in the rhizosphere are the "elders"—they're like a community or a village that not only influences both the roots of the plants in the area but can also determine what actually germinates or grows there. These elders control, manipulate, and alter various chemical components in the soil that the seeds need. Meanwhile, the roots of the germinating seedling send out a buffet of goodies—carbohydrates, proteins, and sugars—through their root hairs, attracting the microorganisms that will aid in its growth and development. The seeds give the elders a new and revised vision of the future, hope for the present, and a reason to keep producing, an exchange that keeps the soil system in balance. It's the waste (the wisdom gained through experiences) given off by these microorganisms that

YOU ARE SEEN

Being seen for your true self is powerful. The head nod is an existential carry-on when traveling as a Black male on this planet, a universal greeting among the melanated males. I've used it on three continents, acknowledging that I see you, and you see me. In the Zulu language and culture of South Africa, one of the ways people greet one another is not hello, but *sawubona*, which means "I see you" or more deeply, "I acknowledge your presence and your existence." The response back can be *yebo, sawubona*, which translates to "Yes, I see you, acknowledge you and your existence, too," or *shiboka*, which translates into "I exist, I am here for you." A greeting that truly empowers, honors, acknowledges, and respects those to whom it's spoken. The English word *respect* is derived from the Latin "to see clearly," *respicere*. Whether in a store, a classroom, in an office hallway, an elevator, a church, an airport, on the banks of the Dead Sea, or in the foothills of the Himalayas, the head nod means "I see you, I acknowledge your existence, I respect you as a person and I see you with dignity, pride, and appreciation."

Sawubona, Mende, Mbundu, Igbo, Temne, Mandinka, Fulani, Bakongo, Fon, Wolof, Chamba, Zulu, Xhosa, Ndebele.

allow the root hairs to absorb critical nutrients for the plant. It's a miracle of give-and-take between the plant and all its helpers and providers. Much of this process, very much like what happens in outer space, is unexplainable to the human mind, but as above, so below. There is a multitude of universes, a multiverse under our feet.

I remember little gems from people such as the incomparable Brother Bey—entrepreneur, radio host, and health advocate—who would start most of his statements and his radio show by saying, "The day is pregnant with possibilities." The elder Baba Tarik Oduno, nearly seven feet tall, would infuse into every space he entered, "There is no culture without agriculture," whether it be at a United Negro Improvement Association meeting or at Malcolm X Park during the drumming circle in NW Washington, DC, or down in Capitol Heights, Maryland, at Everlasting Life restaurant. Every Monday evening at Howard University, psychiatrist Dr. Frances Cress Welsing would sow psychological seeds to all who would listen. I was fortunate to catch Dr. Jawanza Kunjufu as a guest speaker in Silver Spring, Maryland—his *Countering the Conspiracy to Destroy Black Boys* book series guided my adolescence. If you were lucky enough, you could catch a master like Baba Dick Gregory, the writer, activist, and comedian, walking down U Street or Georgia Avenue. He was a man whose life and presence set a target for me as to how tall my tree needed to grow.

Sawubona, Baba Dick Gregory, Baba Brother Bey, Mama Dr. Frances Cress Welsing, Baba Tarik Oduno, Baba Jawanza Kunjufu, Baba Dr. Brimmy Olaghere.

This rhizosphere is the equivalent of the community and environment you are born into. The rhizosphere has both helpful and harmful elements. It can be home to negative elements such as root-eating nematodes, late blight, soil pathogens, heavy metals, and toxic substances, which can be detrimental to root development and plant growth, potentially killing the plant or creating disease before the plant has barely begun to grow. A potato blight, or a crack cocaine epidemic in a community, are both results of a poor environment and are generally brought in by others, as these pathogens don't thrive or spread naturally without other outside factors in place. They can negatively affect the seed but don't define the seed or its ability to grow and thrive when these factors aren't present.

My presence, questions, comments, and family history were my exudates—the elements I offered to the environment around me. And in turn I was able to interact with so many amazing people, personalities, concepts, institutions, places, events, and spirits that fed me and my growth. Just like in the rhizosphere of a plant's roots, humans interact and engage with billions of other entities as our seeds start to send out roots. This rooting process in our environment can be and sometimes should be uncomfortable, because many of us find ourselves in places we weren't meant to be planted. Some of us African Americans, Afro Latinos, and Afro Caribbeans are descended from people who were kidnapped and enslaved in hostile environments that remain unconducive for our growth to this day. Despite the challenges, we grow and develop, morphing from seed into the fullness of our selves.

Mr. Sam Cooke reminded us that a change has got to come. I'm not sure if every seed knows or thinks this, or if a seed can identify that it's been planted or placed in an untenable place, a situation that's not ideal. Does a seed know and lean into its nature, and ultimately grow itself into the change, into the plant species that created it?

GERMINATION

Seeds can't help but germinate when the conditions are right. That doesn't mean those conditions have to be ideal; just good enough. If there's darkness and water, and the temperature is close enough to ideal, that seed will sprout. Seeds ultimately germinate in the dark. In the darkness of whatever situation you may find yourself, if a seed is there, it's most likely going to germinate—watered by physical or emotional tears. I can hear Benard Ighner's velvet, soulful, and soul-filled baritone voice lamenting that nothing stays the same in his 1974 ballad "Everything Must Change." Seeds have to change.

For me, the turning point when the seeds of my purpose began to germinate was in 2003. I experienced a spiritual change right as my great-grandmother Grandma Lizzie, Mrs. Elizabeth White, was transitioning out of this human form, and I became an African Hebrew Israelite. I started to travel back and forth between Israel and DC. In Israel, I stayed near an urban kibbutz in the largest African American community outside the United States, where I worked with the minister of agriculture of the Kingdom of YAH, Minister Sar Yadiel Ben Yehudah (he was a minister when I met him, and at the time of this writing is a prince in the community). This was and is the largest organized group of repatriated African Americans in the world: a community of about 3,000 people of African descent living in Dimona, Israel. I worked at the village's community farm, growing mostly familiar European- and Mediterranean-based crops such as brassicas (kale, collards, broccoli, cauliflower), beans, eggplants, and Swiss chard (referred to as Hebrew chard by Prince Yadiel). The farm utilized their version of a CSA (community-supported agriculture) project, using organic and sustainable growing principles and drip irrigation.

As the seeds in me began to germinate, I found myself in touch with seeds in a literal way for the first time in my life. In my annual visits to Israel, Prince Yadiel tasked me with sourcing and picking up heat-tolerant varieties of seeds for the community farm. Through his words, his interest, and his exuberance about divine agriculture, Prince Yadiel watered the seeds that had been planted in me by my biological father, Big Mike (or as he likes to refer to himself when I'm around, The Michael Carter Sr.—the original). Prince Yadiel provided wisdom and references about the significance of divine agriculture in the Bible, and its relationships to the larger creation. The ntintan was being shown to me: the interconnection between agriculture and everything else. Sawubona Prince Yadiel.

The community taught me about the plethora of biblical laws that speak to agriculture. Right away in Genesis, Adam is placed in a garden—a.k.a., a farm—and he's given instructions to till and keep and garden. So the first occupation that the Creators ever created was a "farmer." That gave me a different outlook on agriculture—beforehand, I had a general idea that farming was dirty work, hard work, without the value and respect of being a banker, doctor, business manager, or even fashion designer. Farming was, in my value system, very low. But if growing crops had such importance in the Bible, if it meant reverence and appreciation for nature, then maybe farming should be valued more highly. I began to think about the metaphysical and the spiritual side of farming, and about whether I might value becoming a farmer myself.

The seeds were starting to germinate. And the right environment was providing what those seeds needed to grow.

The phrasing of the griot from my family reunion—"a looooooooooong time ago"—was another seed that soon found the right conditions for germination. Prince Dr. Shaleahk Ben Yehudah, one of the founding princes of the African Hebrew Israelite community, used to use the phrase "I love the land" on his sacred visitation tours of Israel. He usually followed this by calling everyone in earshot "babies" in a raspy but bellowing voice, exuding confidence and security in all the words to follow. He would recount history about "the land"—the land of Israel, specifically the lands of Judea. I never met the warrior-scholar; he transitioned from complications of a stroke in 2004, as I was just getting to know the community. However, I became good friends with his son, Ahk Abshalom Ben Shaleahk, and he'd always say I had a spirit of learning similar to his abba (father), and a love or an appreciation for the land.

As it became my turn to teach and speak, the griot's words merged with those words of Prince Dr. Shaleahk. I would say, "I looooooooooooooooo

OOOOOOOOoooove the land," making the statement in four or five different countries, on two continents, and in three or four languages, hoping to connect with those I was speaking to, and hoping to speak to the land itself, in a vibration familiar to each particular place. My love of the land, my growing respect for farming, and my calling as a teacher and leader were all coming together.

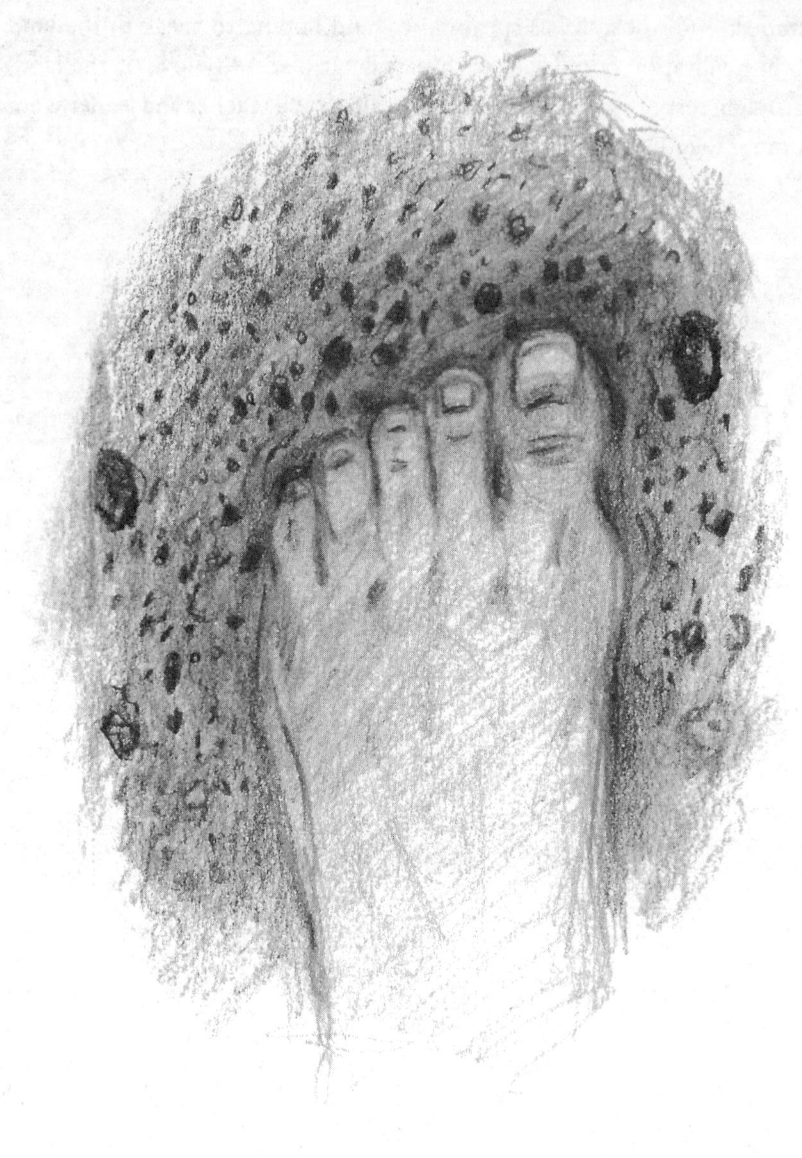

CHAPTER 2

SOIL

THE SUSTAINING SUBSTRATE

Whenever the soil is rich the people flourish, physically and economically. Wherever the soil is wasted the people are wasted. A poor soil produces only a poor people—poor economically, poor spiritually and intellectually, poor physically.

George Washington Carver[1]

If heaven is in the sky, then it's also in the soil.

There are as many organisms in the soil as stars in the sky. These life forms are so numerous it's hard to grasp.

Soil, an Anglo-Norman French-derived word from the Latin *solium* and *solum*, meaning seat and ground respectively, is the base for most life on this planet. As any agronomist or soil scientist will emphatically emphasize, soil is not dirt. Dirt is what you get on your shoes and pants and in your fingernails after engaging with soil. Where the seed represents the individual, soil represents the communal. Seeds grow best in nurturing soils.

Soil is complex and dynamic, being shaped by weathering, environment, human interaction, atmospheric and alluvial (wind) conditions, and a host of other factors. In my early agricultural classes, my father and my other agricultural teacher, Mr. James White, taught me there are three basic types of soil: silt, sand, and clay. Loam is a mixture of these soils in various proportions, with the more prominent sediment in a particular soil being the lead word in the soil's name: sandy loam, silty loam, or clay loam. There are also peat and chalk soil types found in various parts of the world. Soil is a result of the interaction of organisms with eroded geological layers, and the nature of the soil determines which plants can grow and how well they are rooted. Even when you see plants growing up through cracks in sidewalks, it's the sand, water, and limestone from the Portland cement in the sidewalk that is feeding and sustaining that plant.

According to the Virginia Department of Conservation and Recreation, Virginia alone has about 500 different soils. The state soil of Virginia (yes, there are state soils for every state in America), is the Pamunkey soil located around the James River. This soil sustained the Pamunkey Nation for many hundreds of years, and allowed the Pamunkey and other nations to provide First Nation welfare to the initial illegal immigrant settlers of North America in the early 1600s. These are some of the soils my ancestors worked, nurtured, and stewarded before the first Europeans arrived in America.

Our family farm resides on Tatum soil, a deep burnt orange clay, or heavy clay as described by the USDA. In some agronomic circles it could be considered bad or poor soil. It's no coincidence that my great-great-grandparents were afforded the opportunity to purchase the land we have, situated off Tatum Road. African Americans in Orange County, throughout Virginia, and scattered across the Southern agricultural belt or Black Belt were sold the soils that were deemed poor and less desirable. But like with everything we have faced in this country, we have turned poor soils and undesirable locations into prosperous soils and productive places of love, purpose, and tradition.

It's stated quite frequently that prostitution is the oldest profession on Earth, but I emphatically state that the oldest profession on Earth is that of farmer and earth caretaker. We were innately soil men and soil women before we were soul men and soul women. From the rising and falling of the Nile during the first signs of Kemetic (ancient Egyptian) civilization, to the gardens of Mesopotamia, to the sprawling civilizations, communities, and agricultural enclaves of thriving abundant nations and empires, we have a great relationship with our first mother, Mother Earth.

The physical layers, or horizons, of soil start from bedrock (layer R) and move up to the parent rock (layer C), then to the subsoil (layer B), the eluviation (layer E), topsoil (layer A) and the tippity-top, the organic layer (layer O). Soil layers, or horizons, are created from the interactions of history, place, and the environment. Pedogenesis in agronomic terms is "soil formation" in

MOTHERS OF THE EARTH

Cultures the world over have always venerated Earth as a woman, as the woman, the most divine but the most feminine in every aspect of her nature and attributes: soft and subtle, highly responsive to good and pleasing treatment, yet highly vengeful, vindictive, and even spiteful if not treated right, fairly, or justly. We are reminded of the multitude of reverences various cultures have had for this motherly spirit and ethos—for example, the Greek goddess Gaia and the Roman goddess Ceres. In Africa, as in Asia, Europe, and the Americas, the reverence of the female spirit was no different.

In the Fante area and subculture of the Akan culture, Thursday, Yawoada, is the day you aren't supposed to farm or do any work with your land, to honor Nana Asaase Ya, Mother Earth and the wife of Nyame, the supreme being of the Akan culture.

Before the Niño brothers gave Columbus a maritime "Lyft" that led to the annihilation and destruction of the Taino Nation of the Greater Antilles, Atabey, the Taino ancestral mother of Earth, the waters, and fertility was honored and revered.

layman speak. This process is oddly or coincidentally (but not really odd or coincidental to me) similar to the development, shaping, growing, maturation, erosion, and extinction of people and groups.

Soils can also be classified as a series within a family that in turn is within a subgroup, great group, suborder, and order. The USDA recognizes 12 soil orders, 47 suborders, 185 great groups (some sources say up to 305), 4,500 soil families, and over 14,000 soil series.[2] The Pamunkey soil is an example of a soil series.

Soil is at least as complex and just as misunderstood as we humans are. It reminds me of the last scene in *Men in Black II*, when Agent J opens a locker at a bus station that serves as a portal to other worlds and dimensions. Every time you put a shovel, spade, or tiller in the soil, you are potentially accessing a new dimension or galaxy of soil. Within an eight-ounce cup of soil, there are more living beings than there are people on the planet—communities of bacteria, fungi, nematodes, insects, algae, microarthropods, invertebrates—scores of organisms not seen by the human eye or acknowledged in the human consciousness. But we tend to view soil only as a substrate in agricultural terms—something that holds plants upright, that keeps them stable. And that stability is central for growth. Every time you take a step, that soil is upholding you. Even when you're in water, there's soil upholding the water. It's the crux of earthly existence. Stability is also a principle of Africulture.

It's easy to take soil for granted, since it's always there, but the soil is what feeds people as well as plants, what gives us all the nutrients and sustenance that help us grow into our fullness. As much of humanity around the globe progressed from hunting and gathering to agrarianism, agriculture became our staple as a society, our stability. When farming proved not viable in one place, we moved to where we could grow food consistently for the long term. Our cultures are literally built on and around the soil; soil is a metaphor for the larger environment we live in, but it's also a direct physical connection among people, food, and culture. The origin of the word *culture* is tilling or tending the land—cultivating the soil. We can also cultivate our minds and manners, and we cultivate each other through living together as family or a people. Prior to the colonial era in the United States, food and farming were core foundations of culture. This is still true in many societies where specific crops are essential to life—and that foundation is acknowledged in agricultural festivals that celebrate planting or bountiful harvests.

Soil offers stability from a psychological as well as a physiological perspective. If you want to see a stunted plant, dig up a tree and try replanting it. African Americans' growth has been stunted in the West—we have been moved from our native soil into a soil polluted by race and racism. People of African descent residing in the Western hemisphere have been shaped and molded, even soiled (all my puns are intentional) by environment, trauma, abuse, abrasive laws, and social mores that have repeatedly eroded who we are and derailed our purpose, brilliance, and full contributions to our larger human and earthly family.

Soil scientists tell us that the quality, complexity, richness, fertility, and viability of the soil is based on the two base layers of the soil: horizon R, the bedrock layer, and horizon C, the parent layer. These layers represent the determining characteristics, the traditions, if you will, of the soil. The bedrock is solid and stable, the foundation of a culture. Everything is affected by the strength or weakness of the bedrock and the parent material it creates. It's the same as with a family: The more that you know about the bedrock, the more stable the parent material will be, and the subsequent generations (soil horizons) will have a higher likelihood of being productive and fertile.

This principle of home is formed by all the microorganisms and microelements that make up the structure we call soil, and by all the foods that that soil can grow. When imperial opportunistic kidnappers took us away from our home, when they destroyed our home by taking us away, all we had was what we could remember from that soil. For me, this means that my bedrock and parent material were eroded. I was fed only from the top two layers—the organic layer and the topsoil. As an adolescent, I became hyperfocused, I would say maniacal, about wanting to reconnect to my deeper layers. As my curiosity increased, it inspired more determination to discover who I was, what soil I came from, and what that soil had made me.

I hearken your attention back to 1977's *Star Wars: Episode IV – A New Hope*, when the Death Star blows up the planet Alderaan, or the 2009 film *Star Trek*, when the planet Romulus explodes because of a supernova. The characters who come from those planets experience anger and frustration, but also a sense of purpose. Their home planet was their soil.

The transatlantic slave trade—with its undercurrent of anti-Black and anti-farmworker bias—was essentially an explosion of our home planet in Africa, and many of us have been trying to return to Alderaan, so to speak, ever since. It's been an obsession that permeates the very cells in my being, consistently attempting to make connections with African or ancestral consciousness and culture. Infuriated at what was taken from me, making do with what's been pieced together to provide a sense of community, home, and familiarity. For me, one way to connect is through the ancestral experience of being nourished by the same foods that grow in the soils of Africa; another is by learning about the culture that sprang out of those soils.

LOOKING BACK

When I moved to Ghana, it was a major step toward realizing a goal I had probably had, but couldn't see, ever since adolescence. In 1987 and 1988, I was in Miss Miller's fifth grade class, and in every writing assignment I was tasked with, I wrote about something related to Africa and African culture.

Sankofa adinkra symbol. *Photo by bagaball / Wikimedia Commons.*

That summer was a high time of African consciousness in hip-hop music and in my life. My cousins would wear leather medallions associated with Africa, emblazoned with an X for Malcolm X, or adinkra symbols representing parables and principles, like Sankofa. Sankofa is a bird that turns its neck back toward its tail, as if trying to take a bean or rock or seed off its back. "To retrieve" is the literal translation of *Sankofa*, noted in the proverb, "Sɛwo werɛ fi na wosankofa a yenkyi," which means "It's not taboo to go back and fetch what you forgot or left behind."[3] Sankofa has become the symbol for people of African descent repatriating back to Africa either as tourists or residents.

And in that time in the 1980s, hip-hop artists were expressing these ideas in their music. Public Enemy, De La Soul, and X Clan were all talking about returning to Africa, African principles, and African pride. My first intro to Malcolm X was through Public Enemy and the X hats people were wearing in their videos, not through classroom schooling. It was a positive way to learn about him. I was ten, and it incited in me a pan-African reality that engulfed the rest of my life. I'd never heard of our direct connections to Africa. I just anecdotally knew, because of our skin, we must be from Africa.

In the fall of 1989, I checked out a book at the local library called *Roots*. The way Dr. Alex Haley described how the children in Africa interacted reminded me of how my friends and I interacted. Throwing insults to each other's mothers as a playful thing—that's what we did on the bus every day. I identified with the characters in that epic book, and it was probably the first time I placed *myself* into a book as a character. It was much harder to do that with a Judy Blume or Encyclopedia Brown book, which made up the majority of my literary consumption at the time.

Tracing roots, as in family histories, leads us back to the soil that nurtured those roots. *Roots* felt like a path home. When I read the scene of arrival into the United States at Spotsylvania Courthouse, Virginia, it dawned on me that the story took place in Spotsylvania County, right next to where I lived in Caroline County. I pondered, a bit embarrassed: I had lived next to this history for years, yet I had no idea. *Roots* was the first time an African American got to share their story from their vantage point to such a large readership, as well as the first time on mainstream television that a depiction of slavery and its visceral atrocities were shared in prime time to an American audience.

Later, as an adult, when I would sit at the feet of my mentor, Dr. Brimmy Olaghere, he shared with me that he was one of the individuals who aided Dr. Alex Haley to trace his roots. In the same kitchen where I would visit Dr. Olaghere, he had set down a palm nut soup and a peanut soup, decades before, in front of Dr. Haley, saying "Whatever one you like is the one that's closest to where you are from." He was talking about genetic memory, food cells, and soil. Well, from Dr. Olaghere's account, Dr. Haley devoured the peanut soup. The home of the groundnut in Africa is the Senegal-Gambia area, and this is where his roots start from. We haven't lost everything in our genetic memory. Certain foods still speak the frequency of our motherland and our ancestral homes. They resonate with the soil of the soul.

GROUNDING

We find different ways to find home. For a lot of us, when we dance, we take off our shoes. Having your feet on the soil encourages and inspires you to move and recharges your spirit.

This practice of working, walking, or dancing barefoot is referred to in various New Age and Internet Age terminologies as grounding. Grounding is when you make direct contact with the surface of the Earth, most popularly by standing, running, or walking barefoot on a natural surface. The infiltration of electrons into your body is said to coat red blood cells in your body, reducing coagulation, allowing blood to flow more easily through the veins and arteries, and reducing blood pressure. The soles of our shoes prevent us from connecting with our source directly, our dear Mother Earth, from the surface down to bedrock.

Interestingly, one of the things that emerged when I later reconnected with the soil of Africa was a different relationship to anger. After being in the African Hebrew Israelite community for a few years, my mind was set on leaving America and living outside of Western dominated culture. I was "quitting America," in the same vein that activist, professor of law, and founder of TransAfrica, Mr. Randall Robinson, wrote in his 2004 book of the same phrase. During my first visits to Ghana in late 2006, the radiance of the people, the warmth of the weather, and the wholeness of the foods planted a seed in the soil of my soul that indicated this would be my new home. I was surprised to discover that in Ghana I could truly show my emotions, especially anger and frustration. For decades I had suppressed these feelings, for my safety and the safety of others, cultivating the perception that I was always cool, calm, collected, and composed. In my past and present American life, rarely have I raised my voice in public or gotten into an active argument with someone for any prolonged period (except for a family member or three at the farm).

But once I moved to Ghana in 2012, I threw my emotional inhibitions in the Atlantic Ocean. I noticed how Ghanaians seemed to argue frequently, seeking to get their point across or to be heard in a disagreement. The arguments rarely escalated past the raising of voices, the pleading or sharing of your perspective to someone else; it would come to some sort of resolution, even if it was to agree to disagree, and then you go on about your way. And so I found myself fussing with taxi drivers and arguing with police officers and utility workers whom I felt were trying to shake me down for money. On one occasion, when confronting a water utility worker who had cut off my water, I picked up a pipe or machete, to express my displeasure with their actions. Totally out of character for me, totally. But on that soil, in that environment, my frustration or anger wasn't interpreted by others as a threat—or a reason to deprive me of my life. For the first time in my life, I felt I had a First Amendment, a true freedom of expression that I didn't have to suppress. I haven't had that feeling again since my return back to the land of my birth. It's baffling to think that though I'm an eleventh-or-so-generation American, the only time I have felt comfortable enough to exercise my right to free speech was in a West African nation three and a half decades after my birth.

Ghana officially made me a grounded tree hugger. When I found myself in an angry mood, I'd walk outside barefoot and hug the coconut or acacia tree in my backyard. Almost instantaneously that stress, anxiety, or angst left my body. After half a minute to a minute of deep breathing, resting my head in the bosom of the tree, the caramel, coffee-with-two-creamers complexioned Incredible Hulk transformed back into Bruce "Michael Carter Jr." Banner. This became my conflict mediation strategy for my sons—anytime they got highly angry, they earned tree time. Take off your shoes and hug a tree.

Like many kinds of plants, trees release phytoncides, which are airborne, allelochemical, antimicrobial compounds that aid plants in protecting themselves from harm, in most cases, internal harm. Humans breathe in these plant-based bioactive compounds, which have been observed to lower anxiety, reduce cortisol levels, and strengthen natural killer (NK) cells (which are like a missile defense system for our immunity). Grounding seems to make you closer to yourself and can give you a feeling of home. This was the soil that was being developed around me in Ghana, an environment of connection and memory.

TRANSPLANTATION

As above, so below: Soil also constitutes our bodies. We are made in the image of the earth, with all the proportions of the earth. Our bodies are mostly water, just like the surface of Earth. We're also made up of an abundance of minerals and gases. Oxygen is abundant in both our bodies and the earth, and we contain about the same proportion of calcium as the earth does. In my spiritual philosophies, I was taught that the human was a microcosm of the earth, containing proportionate amounts of the micro and macro minerals that are contained in the earth. Soil is intrinsic to us. We're not just drawn to it, but we come from it. The benediction of "ashes to ashes, dust to dust" at Christian funeral services expresses the idea that we came from the dust.

Afar min ha'aretz or *afar min ha'adamah* are the Hebrew transliterations I learned for Genesis 2:7. "And YAH formed man from the dust of the Earth." I was taught that *afar* could mean minerals, dust, or pulverized particles that make up the soil. This passage conveys connection with the soil and assists us in seeing ourselves as part of the soil, part of our eternal mother.

When you transplant a tree, its ability to adjust depends on the age of that tree and how tightly it's involved in the soil in terms of the greater culture. Is it an old tree, new tree, sapling, seedling? Was all the soil around it taken up with the roots, or are the roots exposed to the air and elements? If the tree is older than a sapling, it'll endure tremendous transplant shock. That shock—the loss of roots, in addition to being placed in a new soil environment—will sometimes be so traumatic that it will kill that plant. That plant is telling us, "I'm too old to be moving like this." It's the plant version of trying to teach an old dog new tricks.

The tight connection between soil and humans means that, just like plants, we can be damaged or killed if our relationship to the soil is severed. The kidnapping of African people out of our soil required re-rooting in a foreign land. As the people rooted into the new soil and new environment, they had to adjust to the cooler climate, the difference in sun exposure and intensity, and the much harsher social climate.

During the transatlantic slave raids and human trafficking, the slavers purposefully selected younger individuals. Older individuals weren't going to adjust to the new environment willingly; they were more apt to die than to fight to live on. The presumed average age of a person taken into slavery was between twelve and twenty-five. Many young people, even though taken against their will, were strong enough to deal with this shock and had enough life ahead of them to not want to give it up. However, some chose the sea. We give a moment of silence and respect for those who made the ultimate sacrifice and died at the bottom of the ocean rather than live on their knees. Sawubona to those who chose freedom in the sea.

With transplant shock, some plants wilt. Some may grow but are more susceptible to diseases and pest pressures; they probably won't make fruit. In human terms, they'll experience some level of depression or anxiety. They'll be there, but they won't perform to their full potential. Others, such as saplings, may seem to survive transplanting without too much harm because although their taproots have formed, they hadn't rooted fully in place. They know who they are but they haven't experienced or *lived* who they are. They may live a full life and even bear fruit, but the shock or trauma is transferred to their seeds, creating a generational legacy of traumas and shocks.

Individual plants can behave very differently in different soils, and this goes for plant species, as well. Horse nettle, a common weed at the farm, and a plant called gboma that grows in Ghana, are both in the nightshade family. They look similar, except horse nettle has little painful spikes on the leaves, branches, and stems. (See page 6 in the color insert.) There's no gboma in the United States, and I didn't see any horse nettle in Ghana. It's my theory that gboma seeds were brought to North America, possibly during the trans-atlantic slave trade but maybe even before that, when North America and Africa were one. The connection between Africa and early America goes much further than Colombus's nautical "Uber" in 1492. Some 200 million years ago, what's now the Southern US was once physically connected to present-day Morocco, Mauritania, Gambia, and Senegal.

To survive in the conditions of North America, the plants that were gboma developed thorns, and so were given the name of horse nettle. The plant has been geographically modified, much like the African who resides in

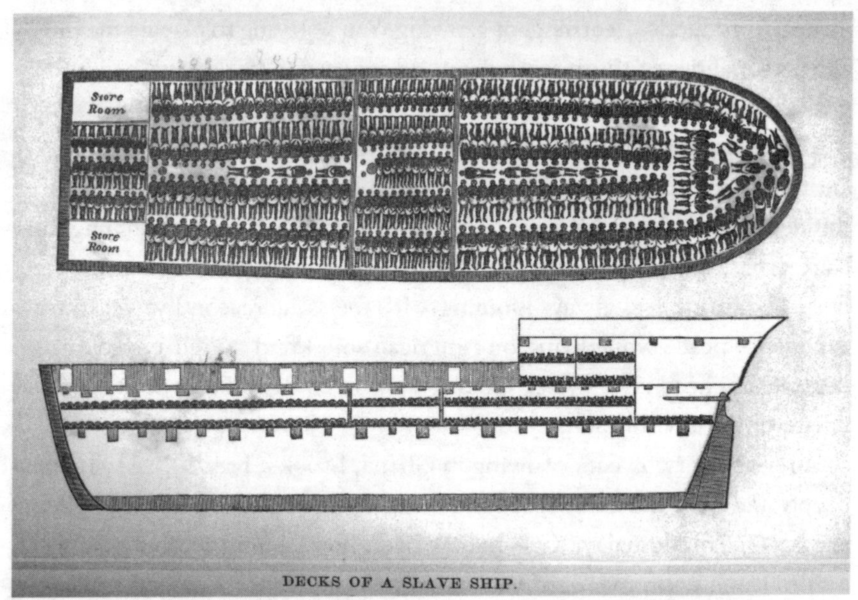

Slave ship. *Illustration by William O. Blake / Wikimedia Commons.*

America—adapted to the landscape of this new, often strange and off-putting social environment.

When my family and I first moved to West Africa, my older stepson was sixteen. He went there with what I called the DC mean mug, or ice grill. He was constantly looking mean or angry at people, which is the norm in certain communities; I wore the same scowl when I was a teen. It's used as a form of protection or as the first tool of intimidation, very much like the thorns on horse nettle. In Virgina, we called it muggin', ice grill, grilling, or mean mug. When he got to Ghana, walking around, he still had that mug on. That was an air of protection he maintained, not knowing what the new environment would hold for him.

After the first week, though, the smiles started. He realized he didn't have to be overly defensive or protective in Africa the way he had to be in Washington, DC; he didn't need those thorns anymore. It was a social awakening: Nothing in this environment was trying to attack him, consume him, or hurt him in any intentional way. Being around so many other gboma plants, he started to become a thornless gboma, too, allowing the cultural geography to shape his physiology and mentality. There is a freedom to realize no one's after you, and now you can channel that energy to something else. When you're trying to protect yourself, your mind is on watching your back, not on being productive, learning, or growing. You're trying to placate the minds and personalities of those around you, as opposed to dealing with your own growth and development.

Enslaved African Americans were constantly on defense—they didn't know what attitude the slave master would have (and in a way, they shared this defensive reality with slave owners, who were constantly watching their back, front, and sides, as the retribution for stealing someone's time, freedom, and future was always looming) . It's the same reason I've yet to truly experience peace while living on American soil. I find myself constantly on defense, my head on a swivel, looking to protect, fight, flee, revolt, or fit in, whatever the environment needs.

After about two years of living in Ghana, I took a breath . . . I breathed deeply and fully, the air of freedom, of peace, of relaxation, of joy, of being at ease. One random day (it might have been after cashing my first paycheck), a calm came upon me. I smiled—not a fake smile, not a staged smile, not a mask in the shape of a smile. I was able to feel at home. Maybe not my geographical home, but the home that soothed my soul. The home that Stephanie Mills sang about in the movie *The Wiz*, the resting place that provides serenity, peace, and quiet.

When the raindrops fell during dry season, or when we didn't have access to running water for a few months, it had a new meaning. In this season, Michael Carter Jr., the person, the plant, the purpose, was reborn, still struggling with straddling the *place* of home with the *feeling* of home, but from a different vantage point. I wanted to keep on feeling the ease and freedom of being in Ghana but still have my family and the friends I'd grown up with close by me. Or I wanted to be able to return to Virginia but hold on to this feeling of peace, well-being, and serenity that had evaded me all my life. Being potted in West African soil for just a few years shifted and changed me in a highly positive way. The soil can have that effect, but also just the opposite effect. A few years later, I sojourned back to America, and as though I'd taken a wicked magic potion, I felt those thorns, briars, and bitterness creep back in. Once again I was geographically modified. America has ignored the blight in its soil and stated emphatically, through its actions, that it needs no remediation plan.

SEVEN GENERATIONS

Bahamian minister Dr. Myles Munroe posited, "When the purpose of something is not known, abuse is inevitable." This quote speaks volumes, and to be more precise, it refers to when the purpose of something is not known in a Eurocentric, counting-and-measuring sense. Our treatment of soil shows how our worldview has marginalized or undervalued its contributions. America's Dust Bowl history is a perfect example.

Soil degradation, abuse, mismanagement, and lack of stewardship, coupled with meteorological factors, created the Dust Bowl of the 1930s. The first and most underappreciated abuse was the act of running the original soil managers (the Indigenous nations) off the land, which they had stewarded for hundreds if not thousands of years. The agricultural prowess and wisdom of nations including the Caddo, Osage, Omaha, Yankton, Santee, Iowa, Kaw, Arikara, Nakota, Missouria, Mandan, Ponca, Pawnee, Quapaw, and Hidatsa, had long assisted in maintaining upward of six feet of topsoil that anchored grasslands and prairielands. This proper management worked in harmony with nomadic horseback nations such as Blackfoot, Crow, Cree,

A Dust Bowl storm churning up topsoil. *Photo by George Everett Marsh Jr. / Wikimedia Commons.*

Tonkawa, Sarsi, Cheyenne, Apache, Kiowa, Lakota, and many others that followed the migration of buffalo. Keeping a circle of life flowing for centuries, these farming and environmental practices preserved the land and the water, and provided a vibrant ecologically balanced home for its residents.

In a matter of fifty years, this balance was destroyed. Practices such as crop rotation, land rest, native grassland management, controlled hunting, sacred space acknowledgment, and seventh generation–based decision-making were seen as primitive, unproductive, and unprofitable. Instead, settlers practiced overgrazing and deep plowing that caused massive erosion. Crop yields took precedence over crop selection and crop rotation. The arrogance, combined with ignorance, of the newcomers reduced six feet of topsoil to six inches in just a few decades.

Turtle Island possessed a solid soil foundation, a bedrock, whose plant material and topsoil started to erode long before the Dust Bowl—some 440 years ago, with the beginning of colonization. Defining land as a commodity, a personal resource to be bought and sold, rented or leased, with its various components isolated and exploited is a far different perspective than understanding land as sacred and interconnected, communally owned by no one but cared for by everyone. This true bedrock, the foundational

Haudenosaunee directive, mandate, and guiding principle is another principle in common with Africulture: Before you act on a decision, understand how it will impact seven generations to come. The idea is based on total unselfishness and rejection of your immediate desires. This is a principle I've seen exhibited in various parts of Africa, in conservation and preservation practices related to land, water, and the air. As no human created those things, no man has a right to take them away from future generations.

The spirit of the original principle has been steadily mined out of American soil. Colonizers, conquistadores, and slave catchers rarely considered how their treatment of Africans or the land would affect the next seven generations. They followed a different mindset: If it don't make money, it don't make sense / cents. Or, as the Wu-Tang Clan stated so emphatically and poetically in 1993, "Cash Rules Everything Around Me . . . (C.R.E.A.M.)." For Indigenous people, C.R.E.A.M. would have been *Community* Ruled Everything Around Me (. . . really "Them," not "Me" . . . but C.R.E.A.T. just would not have been a hit song and motto).

Yet, there are places where the original principle has survived. When my great-great-grandparents Mr. Jefferson Davis and Mrs. Catherine Walker Shirley purchased their land, between the red oak tree, the beech nut tree, the bay white oak tree, and the creek branch, they didn't know my grandfather, my father, or myself. Yet they foresaw how it would benefit us, that land ownership was an opportunity for freedom, independence, and *inter*dependence for future generations. My children are the sixth generation to benefit from this land—this soil my family has stewarded. And many families of all ethnicities hold the same vision of land ownership as my great-great-grandparents.

DISTURBANCE AND RE-ROOTING

Both plants and people can adapt to a range of contrasting environments, but plants and people sometimes face soil or community environments so toxic and disruptive that they are heavily affected, despite their best adaptive methods. Mutations form within these structures in an attempt to adapt or resist an environment that does not affirm life.

Farmers use cover crops to help maintain the soil; those crops keep nutrients cycling and protect the ground from erosion. But the original cover crops, first and foremost, are trees. There's not a lot of erosion in the forest. A forest is an organized and divinely intentioned environment for soil growth, creating soil that allows plants to grow to their full potential. When you start cutting down trees, the first thing that happens is erosion.

As we cut down our trees—our elders, our institutions, our culture, our songs, our languages—we're destroying our soil. When rain comes to nurture the soil, and those "trees" aren't in place, those principles are eroded, washed away into the mainstream pop culture milieu of this American melting pot.

Black people in America have been victims of uprooting, first by being taken from our homelands and then by the constant tillage of American history. Every time we get settled, there's another disruption in the social system that doesn't allow us to fulfill our full reality or purpose.

I have been promoting no-till methods of farming for years. I explain it to people in this manner: No-till agriculture keeps microorganisms in their places, so they can focus on providing nutrients such as iron and calcium. If your soil needs some amendments, you can provide that, and you can plant cover crops that give the soil more diverse nutrients. But when you till your garden or your field, you are sending a hurricane through the top six inches of the soil. Afterward, the community in the soil has to be rebuilt. It's just like after a hurricane strikes a town or city: there's no school, no work, you're out there fixing all the damage that hurricane has created, which makes you unproductive as a person. That report is not going in, that football game is not going to happen, that wedding is not going to take place at the scheduled time. The next two to five months, you're just rebuilding. The same thing happens in the soil after the tiller goes through. Those microorganisms and other elements of the soil are just trying to get back to a healthy productive state so they can provide proper nutrition for the plants.

Now, imagine as soon as you get finished with rebuilding, another hurricane comes and does it again. When are you ever productive? This is what the soil community is dealing with when we till season after season. This is why vegetables and fruits from the rainforest are some of the healthiest for us to eat: they are growing in soil that's not disrupted, where nutrients are available from many sources. The roots of rainforest trees have been growing undisturbed for generations. The bacteria, protozoa, and fungi in that rhizosphere know what they're supposed to do. They're professionals; they've taught their children and grandchildren. In other environments, we've destroyed our fungal network, so the infrastructure of our nutrition system is destroyed. Instead of the soil sending sophisticated nutritional messages throughout its own structure and into plant roots, it's stuck using the equivalent of Morse code.

It's interesting to note that some of the staple European crops like kale, broccoli, and collards, don't depend on fungal networks to grow. Instead, they depend on having a consistent supply of nitrogen. But other plants, including most African-derived food crops, need that fungal network to thrive. Relationships and community are key.

STARVING THE SOIL

When soils fall victim to contaminants—man-made or natural elements that become toxic to living organisms—they have to be sanitized or effectively remediated to eliminate the pathogen permanently. It's a costly, inconvenient, but necessary action. Living with the pathogen in the soil puts every plant at risk, because the longer the pathogen is there, the more it adapts to infect plants whose roots or gene structures were previously immune to it. Living with pathogens is a topic I explore in depth in chapter 7.

Soils also struggle when they are not provided with the organic matter and nutrients needed to support the thriving of the living soil ecosystem. And we can see a parallel here with the unequal treatment of Black farmers in the United States dating back more than 150 years. The 1860s saw passage of the Morrill Land Grant College Act, which set aside federal lands to create colleges to "benefit the agricultural and mechanical arts."[4] These land-grant schools might as well be called land grab schools, since the land was largely in Western territories stolen from Indigenous nations. These schools grew to be a key part of the American educational system, a national network of state colleges and universities, that offered top-notch agricultural education—but as is customary in apartheid America, initially these institutions were largely made available only to white males, with very few exceptions. It wasn't until 1890 that the Second Morrill Act established agricultural schools specifically for Black students. Tribal colleges and universities weren't given land-grant status until 1994.

In 1914, the Smith-Lever Act created a nationalized extension service for American farmers—but it was a service that enacted yet another injustice against Black farmers. Intended to bring the latest agricultural knowledge to rural farmers by "extending" the land-grant universities into rural communities, the Cooperative Extension Service was highly influential—still today, nearly every county in the US has an extension office that runs 4-H programs and distributes informational and consultative agricultural support to farmers and landowners. This concept of support, what's known presently as extension services that provide technical assistance, had been developed by Dr. George Washington Carver, the nation-saving scientist who taught for decades at the Tuskegee Institute in Alabama (more on him in chapter 5), but the government gave no financial assistance to the historically Black agricultural schools for running these programs. It wasn't until 1972 that the HBCUs received federal funding for extension services, and they were underfunded and run by white administrators. To this day, white and Black extension services are funded unequally.

A 2023 report from the Century Foundation detailed the funding inequities between the original land-grant universities (the 1862 institutions) and the "1890s"—the historically Black schools, which together enroll about 117,000 full-time students, 75 percent of them Black.[5]

- The 1862 institutions have endowments that are six times those of the 1890s.
- In the 2019–2020 school year, the original land-grant schools garnered $2 billion more in revenue through federal, state, and institutional support than the 1890s.
- The original land-grant universities are able to spend three times as much on research per full-time equivalent student.
- After the initial eighty-year period in which the HBCUs received no money for research or extension services, they have endured continuing shortfalls in the money they receive versus that promised by Congress—a difference of $436 million between fiscal years 2008 and 2022.
- While fully funding white land-grant schools, the states have failed to provide matching funds to the HBCUs, to the tune of $200 million between the 2011 and 2022 fiscal years.

In 2014, Congress publicly acknowledged the discrimination practices that persisted from 1914 to 1972, with this statement by Dr. L. Washington Lyons to Congress:

> The land-grant system was created by the Morrill Acts of 1862 and 1890. The 1862 Morrill Act created a land-grant university in

> each state and the 1890 Morrill Act extended the land-grant status to the Historically Black Land-Grant Universities in the southern states and the border states.
>
> The Smith-Lever Act, as you know, gave rise to the Cooperative Extension Program in 1914, which is a unique partnership between the US Department of Agriculture and the land-grant universities. However, when the Smith-Lever Act was passed in 1914, it created Extension at the 1862 land-grant universities. The Act did not provide funding for the 1890 land-grant universities at that time. In 1972, Congress appropriated the first funding to support Extension at the 1890 land-grant universities, and that gave rise to the support of both 1862 and 1890 land-grant universities sharing the responsibility of implementing Extension programs in the southern states as well as throughout this country.[6]

Unfortunately, in 2025, it looks like any more progress for HBCUs will be pushed back by federal funding cuts. In many cases, the 1890s universities aren't even allowed to call their agricultural and farmers support staff extension agents. I worked with the Virginia State University Small Farm Outreach Program, where they were called small farm outreach agents and, later, program assistants. And they were restricted to working part-time because the program has a limited number of full-time positions (only two when I worked with them). Most of the twenty-five staffers were retired agriculture professionals who didn't require benefits from the job, because they were acquired from their previous occupations. The Commonwealth of Virginia has kept the same line-item budget for the program from 2000 to 2023: a paltry $394,000 for a program that, as of 2022, is to serve over 37,000 small farmers. In contrast, over this same time period, research funding for hybrid striped bass in farm ponds increased by over 100 percent.[7]

Two years after Smith-Lever, the Federal Farm Loan Act of 1916 established what is known as the Farm Credit System, to make it easier for farmers to borrow money at low interest rates, and farmer cooperatives, so that farmers were essentially lending money to each other. It created twelve federal loan banks, financing them with half a million dollars apiece, and numerous local farm loan associations. The Farm Credit System, along with the farmer cooperative system, quickly became a major economic force: By 1919, there were 4,000 farm loan associations and millions of dollars had been loaned to farmers—white farmers, that is.

This era was also the peak of Black land ownership. More Black farmers than ever owned land, yet they were cut off from being able to access cheap loans that would have allowed them to make improvements and build up their businesses. If they wanted to build up their operations, they had to do it through community financing or on their own dime. This was deliberate neglect of the economic soil of the Black farming community, and as a result, individual Black farmers could not thrive to their full purpose. These policies are equivalent to questionable unsustainable soil practices such as over-tilling (leading to erosion in the form of eroding opportunities) and mineral leaching and depletion (in the form of young people leaving the farming community to do something more financially and socially viable). There's also a parallel to creating a soil hardpan, which is a compacted layer below the soil surface created by the destructive nature of tillage and the weight of the tillage machinery. A hardpan is a barrier that doesn't allow nutrients, water, or other resources to permeate through, which limits the growth potential of plants. And these loan policies prevented resources from flowing down to support farmers in communities, slowly choking them of life, opportunity, and hope. Coupled with the exclusionary loan program was a discriminatory buying policy that shut out Black farmers. Because of segregation, there was no opportunity or desire to buy from Blacks. Separate but equal was the order of the day.

Separate but equal also leads me to think about what I call my Band of Brothers—a group of my great-uncles who had all been in the military during World War II, the Korean War, or Vietnam. They all had this unspoken, quiet but stern respect for one another. And these men helped mold me. They were all responsible fathers and caretakers of their families and the land. Yet they were victims of the society of that day which saw Black people as needing to be put in their place, with many being denied the educational and mortgage benefits of the GI Bill.

The educational provisions of the GI Bill read in part:

> TITLE II—CHAPTER IV—EDUCATION OF VETERANS
>
> Any person who served in the active military or naval forces on or after September 16, 1940, and prior to the termination of hostilities in the present war, shall be entitled to vocational

> rehabilitation. . . . The Administrator shall pay to the educational or training institution, for each person enrolled in full time or part time course of education or training, the customary cost of tuition, and such laboratory, library, health, infirmary, and other similar fees as are customarily charged, and may pay for books, supplies, equipment, and other necessary expenses, exclusive of board, lodging, other living expenses, and travel, as are generally required for the successful pursuit and completion of the course by other students in the institution.

And the mortgage assistance provisions:

> TITLE III—LOANS FOR THE PURCHASE OR CONSTRUCTION OF HOMES, FARMS, AND BUSINESS PROPERTY
>
> Any person who shall have served in the active military or naval service of the United States at any time on or after September 16, 1940, and prior to the termination of the present war and who shall have been discharged or released therefrom under conditions other than dishonorable after active service of ninety days or more, or by reason of an injury or disability incurred in service in line of duty, shall be eligible for the benefits of this title. Any such veteran may apply . . . to the Administrator of Veterans' Affairs for the guaranty by the Administrator of not to exceed 50 per centum of a loan or loans for any of the purposes specified.[8]

Yet these benefits largely bypassed Black vets. My grandfather didn't attend college. Instead, he earned a certificate from the George Washington Carver Regional High School in 1953.

Many Black farmers had to remain in agriculture because it was one of the only available, viable occupations that provided a way to support their families legally.

So, all through the years after emancipation and through the twentieth century, Black farmers were laboring in a system, in a soil, that kept their businesses smaller and less profitable than their potential, while their white

Commonwealth of Virginia
Department of Education
This certificate is awarded to
Warren H. Carter Sr.
who has satisfactorily completed 36 months
Institutional On The Farm Training at the
G. W. Carver [illegible] High School
under Public Law 346
Training Objective General Farmer
April 17, 1953
A. C. Washington Instructor
[illegible] Superintendent of Schools
Willie Yager Chairman, County Veterans Training Committee

My grandfather Warren Carter Sr.'s certificate in general farming, the extent of the educational benefits he was awarded through the G.I. Bill for serving in World War II.

counterparts were supported and given opportunities to thrive by the federal government. This is a major part of the wealth gap that we see today between Black and white Americans. The soil these two groups were growing in was not nurtured in nearly the same way.

How do you rebuild the soil? Leave it alone. Take away the impediments that poison it. Give that soil time to repair itself. In regenerative agriculture, you apply a cover crop, use your compost, and use no-till practices. The framework for regeneration, restoration, and healing is "do no harm." You don't cut down the trees; you figure out ways to utilize the trees and incorporate them into your pastures or fields. You recognize they are anchors that support the greater health of the pasture. You provide whatever resources are needed to help that soil repair itself and don't interfere. It's the same with Black farmers and Black communities. When we're left alone, and the deck is not stacked against us, we thrive.

Once an individual has an experience connecting to the soil, once that understanding is created, encouraged, and sustained, they become fraught with desire to have a garden, have a farm, to *grow* something. Working with our hands in the soil releases serotonin and lowers cortisol levels in our bodies.[9] Being with the soil is calming to the mind and spirit; it's therapeutic. Just the smell of the earth makes us feel better. And research studies show that contact with bacteria in the soil can boost our immune systems and make us more resilient to stress.[10] So literally, soil is essential to our mental health and nurturing of our souls.

In the deepest way, we come from the soil, and whatever happens to the soil happens to people, as well.

CHAPTER 3

ROOTS

BENEFITS AND BARRIERS

When the roots are deep there is no reason to fear the wind.

African proverb

The common root, of course, comes out of Africa.

Duke Ellington

The germination of a seed is also the moment when the first root growth occurs. This is a triple-edged experience. At one and the same time, the roots are the destruction of that seed in its original form, but they also represent its plans for the future and connect it with its past. Right away, as those roots start to spread, they encounter the soil and all the positive or negative influences that reside there.

In the same way, as soon as a person is born, they begin to interact with everything around them, the good and the bad. The roots are soaking up influences even as they anchor the growing person, forming a foundation and a connection to the environment that will define the course of the person's whole life. For most of us, family is the biggest and strongest part of what we absorb into the creation of "you." This "you" evolves, shifts, morphs, grows, dies, reinvents itself, suffers, struggles, succeeds. And if all goes well, you find purpose and a sense of happiness, joy, and fulfillment in the process.

Roots are generally classified as one of four main types: taproots, fibrous roots, adventitious roots, and aerial roots. Taproots are long or large main roots that focus on vertical growth as an anchor of a plant. The taproot is generally the first root that grows from the seed. The technical term for that first root is *radicle*. Fibrous roots are numerous dense, thin roots at the base of a plant stem, like the roots of rice, wheat, and most grasses. Adventitious roots form from the stem of the plant (in most cases) and grow more shallowly. Aerial roots are types of adventitious roots that grow above the soil, which serve as support, like the roots around the base of a corn plant, or to absorb nutrients, like on common ivy plants or on the monstera houseplant. Some plants have several types of roots, rooting both deep and wide, providing a sturdy foundation and aiding in their growth and development.

When I was born in the 1970s, my roots grew into a soil environment that placed a high premium on family. The microprinciples (my analogy to microorganisms) found in this soil were respect, education, history, humor, compassion, care, love, and legacy. My family has been connected to the land in Virginia (as many families have been to the land of their origin or settlement) for over three centuries. Landownership, and the self-reliance that landownership provides, has been an extremely important value we've held since at least the end of the Civil War. Along with that comes the responsibility to steward the land and look toward the future, making sure that our descendants will have security and opportunity. By maintaining our ties to the family land, we're maintaining our ability to continue as a family. Our roots anchor us to each other and to this place.

Everybody has roots, but not everybody has the chance to know those roots. And like roots in the soil, the strongest and most nourished roots have a microorganism/microprinciple community that supports them. Many Black Americans have big gaps in their family history because of the lack of records kept during the period of enslavement, many times because we were viewed and valued as livestock, sometimes because of mistrust of census authorities, and sometimes because of complicated family structures. Even though I only learned informally about my family's history when I was a child, I always knew that family ties were important. As an adult I've made it a point to learn as much as I can about my roots—piecing together family trees, learning as many stories as possible, and placing great value on what I discover: the good, the bad, and the ugly. As a blue-passport-carrying American, I also have no problem embellishing or making assumptions according to the information I'm able to gather or access. While growing up, I was saturated with stories of the colonial Founding Fathers and their virtues: George Washington not being able to tell a lie for cutting down a cherry tree, Abraham Lincoln being honest. These stories wove into the fabric of America a certain character, no matter the flaws. I learned about Washington's wooden teeth, learning later how much of a tall tale that was, that Washington had the healthy teeth of people he enslaved removed and made into a denture set for himself. I see the value in the positive branding and public or private image refining of those you honor, love, and respect. We assist in shaping the perception of how others, and especially future generations, see these loved ones, be it good, bad, ugly, forgotten, or any variation of those perceptions. I'm compelled to weave together stories and experiences based on what I know of my family and the community they came out of, as well as the respectable, honorable, and courageous ancestors whom I've read or learned about.

THE ROOTS OF EMPIRE

Ancestral math is a never-ending computation. We start with two parental sets of genes, then four grandparents, eight great-grandparents, sixteen great-great-grandparents, and so on. But for myself, and a lot of Black families, our roots are both known and unknown. I can identify about twenty of my ancestors by name out of my 2,048 ninth great-grandparents. Those I have found names for lived in the mid- to late-1600s and journeyed to America from Scotland and England. The history of my ancestors with pure or majority African or Indigenous roots is not documented nearly as well, and I can trace that history only to the fourth or fifth generation. This stands in contrast with my European roots, where I can go back nine generations, and, for some, even thirteen generations.

My lack of knowledge about my roots mirrors the general lack of knowledge about Africa in American culture. I can recall one of my first introductions to agriculture in school was a lesson about the Nile Valley civilization, but I thought Egypt was in Europe because of the way it was depicted in movies and in history books. The Elizabeth Taylor version of Cleopatra and the Charlton Heston Moses painted my preteen image of who and what Egyptians were. Receiving my first light-up globe in third grade (still have it, by the way), I thought it had made an error. I was scratching my head as to why Egypt was on the outskirts of the African continent and not next to England or Germany.

But the value I ascribed to Egypt and the brilliance of its civilization, founded in agriculture, was positive. Civilizations were and are defined by the agricultural prowess that allows them to build, grow, and thrive. Kemetic dynasties thrived for centuries, and many grains, such as wheat and barley, are depicted on Egyptian temples and pyramids. I learned of papyrus, the reeds that grew along the river banks and were used for making the paper on which Egyptians recorded what they valued. Ancient cows known as aurochs, which multiple civilizations claim to have domesticated, are the genetic foundation for taurine or Bos cattle, the cows many know and eat today, and aurochs are elegantly and honorably depicted in temples along with herdsmen and women. Honey has been found in numerous temples, with the Egyptians being early beekeepers.

You would even think Cheech and Chong, Willie Nelson, Mr. Calvin Broadus (a.k.a. Snoop Dogg), or Wiz Khalifa were of Egyptian ancestry, as there are intricate carvings of marijuana or hemp leaves in Egyptian temples. The Egyptian goddess Seshat, the goddess of wisdom, is directly associated with the hemp/marijuana plant, and an impression of her with the plant is chiseled on the walls at the temple complex at Luxor, Egypt, its completion dates spanning from 1390 to 1213 BCE. Her persona also plays a pivotal role in the creation story of ancient Kemet. These associations, among many others, speak to how highly the Kemetic people and other civilizations valued agriculture and its relationship to humankind. Hemp pollen was found inside the tomb of Ramesses II at the Valley of the Kings. Precolonial and pre–Common Era traces of hemp pollen and hemp residue have been found in Egypt, in Lalibela Cave in Ethiopia, and among the Khoisan people in southern Africa. In *The African Roots of Marijuana*, biogeographer Chris Duvall posits "Cannabis history must be rethought, especially in relation to Africa. Historians have overlooked considerable evidence while repeating rumors."[1] Hemp's roots ethnobotanically trace back to Tibet and parts of China, and in 1753 it was christened *Cannabis sativa* by Carl Linnaeus. Despite this, I lean toward some landrace strains and varieties potentially having African roots, as they possess unique properties not found in other marijuana plants.

The Bible and the Torah are full of agricultural references and metaphors. The authors of those books and early creators and adherents of their belief system made it a point to acknowledge their relationship with nature, starting with the creation story of Genesis. Adam is literally named after the

Earth; in Hebrew, *Ahdahm* comes from the base word *Adamah*, meaning the Earth. He is placed in a garden that stretches from the Nile River in Egypt to the Tigris and Euphrates Rivers of Iraq. I'd go out on a limb and say that's a bit more than a garden—I'd respectfully say that in today's terminology, the USDA would classify it as a farm. This is the first and only job that the Creators, YHWH, YAHWAH, the Elohim, or God established: that of a farmer. The roots are being set for what the importance of this divine occupation should be in the lives of his adherents. Of the 613 laws in the Old Testament, well over 100 specifically relate to agriculture, coupled with a few hundred agricultural allegories and metaphors.

Agriculture in Africa and throughout the world has formed the roots of empire. The Mali Empire, probably the most well-known of the African empires, is remembered for its ruler Mansa Musa and his Hajj to Mecca in 1324 CE. The empire is legendary more for its gold than its agricultural feats, but consider: It's recorded that over 60,000 people traveled by foot in this mass caravan to Mecca, over 2,000 miles each way, through a desert, long before the existence of highways or rest stops.

His Excellency Ben Ammi Ben-Israel fervently and passionately stated often to his community: "I don't want to be a great individual, I want to be a great individual who is a part of a great people." Africulture has adopted this principle of collectivism and communalism from Mansa Musa. This principle is consistent with the ethos of the Mali Empire, because vast amounts of gold were distributed during his reign. The gold was simply a marvelous byproduct of an amazing people and civilization. The richest man in history did not horde his gold; rather he possessed the intelligence, unselfishness, and compassion to share it with his people, and those they interacted with in their sojourns.

How did they survive? It's unfathomable in my mind, but, of course, they did eat. Agriculturalists and nutritionists must have been key actors of the planning process, feeding both humans and the thousands of animals that were carrying provisions along with thousands of pounds of gold. The amount of gold that Musa brought to Mecca, disrupting gold and financial markets for the next fifty years, is notable. Some accounts state that each person carried four pounds of gold, as pocket change of sorts, and that the eighty or so camels that accompanied them each carried an additional 300 pounds of gold. That's pounds not ounces; it's a tremendous load to carry. It's an even more tremendous load to carry with food. The ability to grow, store, and ration food for that caravan—essentially a walking city—with no granola bars, sodas, bottled waters, potato chips, or plastic-bottled sports drinks, is astonishing.

This agricultural prowess, coupled with gold, is one of the main reasons Africa was explored and ultimately exploited by Europeans. Spain and Portugal sanctioned entrance into African lands, starting in the Canary Islands and slithering their way down the West Africa coast, kidnapping and trafficking brilliance in human form, like a snake that sneaks into a nest and swallows eggs.

The exudates of conquests, rebellion, resistance, terrorism, trauma, pain, death, and upheaval have been feeding my family roots, and the roots of many families like my own, for centuries. But the trouble took place in new forms in the Americas.

In the case of the Carter family, I learned that one of my roots in America is confusing and distorted, but a root nonetheless. This root leads from our ancestors the McIntoshes, who had come to Virginia from Scotland in the 1700s. I assumed they were . . . wait for it . . . bagpipe-playing, kilt-donning, pearly red-haired, pale white European colonists (think "Rowdy" Roddy Piper in WrestleMania I, circa 1985). I would later discover, in a book called *Jacobite Gleanings from State Manuscripts*, a ship manifest that documented a group of McIntoshes who got booted out of Scotland after participating on the wrong side of the Jacobite rebellion of 1745 to overthrow the British crown. The name

10	Alexr Davidson	17	Herdsman	Badenoch	5	5½	Ruddy, slim made.
11	Andrew Edwards	24	Servant	Angus	5	5½	Black, well-made, strong.
12	John Gordon	19	Weaver	Inverness	5	4½	Do. Do.
13	Alexr Goodbrand	30	Carpenter	Bamf	5	6½	Brown, Do.
14	John Grant	40	Labourer	Badenoch	5	4	Black, Do. Do.
15	Alexr Grant	25	Carpenter	Aberdeen	5	6	Brown, Do., lank hair.
16	Josh Hinchcliffe	31	Tallowchandler	York	5	5½	Black, pock-pitted.
17	John Johnson	30	Husbandman	Lanarkshire	5	5	Brown, Do., strong made.
18	David Joiner	20	Labourer	Aberdeen	5	2	Do.
19	Jno Kennedy	32	Do.	Perth	5	9	Black, well made.
20	George Keith	35	Shoemaker	Aberdeen	5	11½	Do. Do.
21	Willm M'Clean	32	Labourer	Inverness	5	3¾	Do., well set, strong, ruddy.
22	Duncan M'Phearson	36	Do.	Do.	5	6½	Thin, pale, and sickly.
23	Angus M'Intosh	26	Do.	Do.	5	6½	Black, very strong made.
24	Peter M'Intosh	34	Do.	Do.	5	4½	Brown, Do., long chin.
25	James M'Phearson	22	Do.	Aberdeen	5	7	Black, lusty, well made.
26	John M'Leod	25	Do.	Inverness	5	4½	Black hair, pale & slender.
27	Duncan Monrow	19	Do.	Do.	5	4	Straight, Do.
28	Alexr M'Leod	18	Do.	Do.	5	5¾	Brown, strong, stares.
29	Charles Morgan	18	Barber	Elgin	5	4	Do., well made.
30	John Murray	30	Weaver	Annandale	5	8	Brown hair, pale.
31	Thomas Ogden	34	Do.	Lancashire	5	5	Do. strong made, lusty.

Ship manifest describing Angus and Peter McIntosh.

Angus McIntosh stood out for me, along with his brother Peter, listed as people deported from England and *described as black*. Mind blown! They were on the ship as indentured servants, having been expelled by the crown of England.

Angus McIntosh arrived in Virginia in 1745. He was listed as a laborer, and the main laboring occupation of that time was agriculture. Therefore I refer to myself as an eleventh-generation farmer. All four of my grandparents' families had been landowners, and they were involved in agriculture in some way.

My seventh great-grandmother Anne was also on that ship with Angus and Peter. Grandma Anne had a child with a Middlesex County servant, an enslaved man named Mr. William Whistler, who was the son of a mulatto woman, my eighth great-grandmother Ms. Mary Whistler. It's still not clear to me whether Mrs. Anne Whistler was considered Black or white. Whether because of differences in race, place of origin, or free or enslaved status, Grandma Anne and Grandpa William's relationship and conception of a child was against the law in the colony of Virginia at the time. Orange County deed book records from April 27, 1758, state that Mrs. Whistler was sentenced as follows: "It is ordered that she pay to the church warden of St. Thomas Parish Five Hundred Pounds of Tobacco or Fifty Shillings or give security for the payment of same at the laying of the next parish levy or on failure thereof the sheriff shall give her twenty lashes on her bare back

My mother's father with his sheep.

at the common whipping post well laid on."[2] She was also ordered to pay a fine to her master or serve him for an extra year after her term of indenture. Miscegenation would continue to be illegal for the next 200-plus years in the commonwealth. (The home of Mr. Arjalon Price, where Grandma Anne was possibly a servant, is now part of the James Madison Museum's Hall of Agriculture & Transportation.)

At times, my complicated ancestral roots leave me feeling stuck and dumbfounded, but they are mine, nonetheless. I have to accept that I have what appears to be a strong taproot, some adventitious and fibrous roots, and some definite aerial roots. My epiphany about the McIntoshes is one of confusion. The challenge in exposing your roots is the risk of vulnerability and uncertainty. My uncertainty displayed itself in my shyness and meekness growing up. I knew I had the potential to feel self-confident, but as I wrestled with my interracial and intercultural identity, I never felt assured enough to be fully open and to be myself around others.

The *Loving v. Virginia* case broke the legal racial barrier in the 1960s. I grew up in the county the Lovings resided in and went to school with one of the Lovings' grandsons, which gave me a close proximity to that history and its impact. Mr. and Mrs. Whistler had broken this barrier in the Carter family two centuries prior, and I was privy to several interracial marriages and relationships that aided in my own existence, including my great-great-grandmother Ms. Alice Whitlow's relationship with my grandfather, Mr. Jacoby Jones, that produced my white-appearing great-grandmother Mrs. Mabel Whitlow (later Fleming). In my very young, naive, and visually impressionable mind, I assumed that when a black person had a child with a white person, Dalmatian-like tendencies could occur. My grandmother Mrs. Sceamer Rose had the skin condition vitiligo; it apparently wasn't explained to me very well. I didn't truly understand it until Mr. Michael Jackson shared that he suffered from this condition, which eroded the melanin in his skin, creating white or cream-colored patches in his skin.

Like many Black Virginians who have a long legacy in this nation, the Carters have a complex family tree. I have grandparents who were white and free, white and indentured, Black and free, Black and enslaved, Black and indentured, mulatto and indentured, as well as Indigenous and every combination in between. Significantly, a 1662 act declared that the child of any enslaved mother in Virginia was also enslaved for life.[3] That ran counter to English tradition, in which a child received status from his or her father, and was meant to discourage white people from partnering with people of African descent.

The roots are always clearer the closer you are to white, and they get much murkier the closer they are to Black and Indigenous.

FREETOWN

Roots stabilize or anchor plants, absorb and consume water, minerals, and other materials, filter out harmful substances, and pump energy throughout the plant. Roots, and most of the plant parts we can see above ground, experience what's called turgor pressure. It's the pressure exerted against cell walls by the fluid in the cells themselves. It can also be referred to as hydrostatic pressure, and it's what makes plants stand upright, tall and erect—or droop when they are lacking access to water. Humans and some animals have our own version of this pressure function. When you pinch or fold up a portion of your skin and it returns back to its normal position, that is a form of turgor pressure.

The xylem and phloem are synchronous escalators in the plant, the xylem always transporting water and minerals up the plant, and the phloem consistently on a downward trajectory, depositing sugars and carbohydrates throughout the plant and then down to the roots. The sugars and carbohydrates become the currency that the roots offer as barter in exchange with the soil for its precious mineral resources that the plant needs to grow.

The process of transpiration (plant breathing) creates a pulling action through the plant, which allows the water to move up against the force of gravity. Roots are purposefully performing scientifically challenging actions to fulfill their roles. They do the best that they can, very much like many elders in our families.

Like roots, our elders have had experiences, encounters, and assaults. They have witnessed or participated in births and deaths, joys and happiness, disappointment and deceit. They have received or been denied justice, ingested or projected discrimination, hate, prejudice, misogyny, misandry, racism. This ebb and flow provide the stability that's needed to maintain ourselves in challenging situations. Just as when drought prevents water from being brought up through the xylem, leading plants to wilt, a lack of access to history and legacy brought about by cutting off our ancestral xylem can leave people starving.

FREETOWN'S CONFEDERATE ROOTS

According to our local historical society, Mr. Robert Ellis, who is one of my fifth-great grandfathers, served as a coachman for Claiborne Rice Mason Jr., a wealthy local physician and Confederate army veteran. It's possible that Mason may have purchased both the land that became Freetown, and the freedom of the Ellises, from Dolley Madison, the widow of US President James Madison, at a time when she was facing financial ruin. As a coachman, my fifth-great grandfather kept gold, silver, and other valuables for Mason, while he was serving in the war, in the false bottom of his wagon. After returning from the war, on the losing side, Mason donated the Freetown land to Mr. Robert Ellis, and the Ellises founded their community there along with about ten other families.

Mason has another historical connection, too. He had railroad-building expertise and became a key builder for the railroads, including the Monopoly-acclaimed Baltimore and Ohio Railroad (B&O). Part of Virginia's lesser-known historical infamy is its convict-leasing policy during and after the Reconstruction era, and it's this system that links Mason to a legendary American folk hero: a nineteen-year-old Black man named Mr. John William Henry.

Mr. Henry—most likely innocently accused and wrongfully convicted of theft in Prince George, Virginia—was leased by Virginia State Penitentiary at Richmond to the Chesapeake and Ohio Railway (C&O) as part of one of Mason's convict-leasing crews.[4] This was a common practice post-Reconstruction, supplying labor for both agriculture and the railroad system, which in turn served the ag industry by transporting commodity crops. Mr. Henry died of silicosis, an inflammatory lung disease caused by breathing in too much crystalline silica dust, during the construction of the Lewis Tunnel or Big Bend Tunnel project at the Virginia–West Virginia border. That's a little different than the way his death has been glorified in American mythology. Mr. John W. Henry is just one of the countless examples of historical asphyxiation, his xylem and phloem choked with the dust of American mountains to build an infrastructure against his will, never receiving any benefits or acknowledgement for it.

Some of my ancestors were part of an early free Black community in Orange County, Virginia, called Freetown. This was a small farming community founded by people who had been formerly enslaved farmers, cooks, chefs, architects, carpenters, engineers, and doctors—but also nation builders. So building a community was light work for them. Mr. Robert "Punch" Ellis, born in about 1790, was possibly a servant in or around Montpelier, the home of the fifth US president, James Madison. Grandpa Punch Ellis acquired the property of Freetown through a unique relationship with a Confederate officer, Claiborne Rice Mason. Personally, I believe this happened before the Civil War—that was why they would have called it Freetown. The hope would have been self-sufficiency and freedom through landownership and farming.

As a youngster, I rode in the car by Freetown, going back and forth to the town of Orange, mostly with my grandmother Mrs. Lucille Carter. I didn't visit Freetown proper until 2023, when I went on an impromptu tour with the owner of Freetown Farms, Mr. Ron Taylor, who is likely my distant cousin. As we were moving around the well-kept Angus cattle farm, dodging cow patties, with Mr. Taylor riding on his ATV, he kept talking about a little pond called Punch's Pond. It wasn't until about my eighth visit that I realized that pond is named after Mr. Robert "Punch" Ellis.

I know little else about Mr. Robert Ellis and the rest of this branch of my family. He died in the 1860s, and he was the great-grandfather of my great-great-grandmother, who eventually purchased the property where Carter Farms is today. So counting me, eight generations of my family in Virginia have been landowners over the last 160 years. As I've said, preparing and providing for seven or more generations after yourself—envisioning future family members you will never even meet—is one of the core principles of Africulture and of stewardship. My life is proof that Mr. Ellis managed to put this into practice immediately after, or even maybe slightly before, emancipation.

ROOTS IN THE LAND

Aunt Dr. Edna Lewis (Kingston), the well-known Grand Dame of Southern Cooking and author, is also from Freetown, and she mentions her Freetown roots in her writings. Her writings describe the Freetown of her childhood in the 1920s and '30s as a very close community where life was based on the seasons and food was always the centerpiece. Meals came mostly from the garden, the barnyard, and the wild fruits and game around the village. Dr. Lewis described her father bringing in honey from wild bees in spring, her mother putting up pickles, relishes, and jellies from homegrown cucumbers and Seckel pears, and herself and her siblings gathering blackberries. She talks about her parents and their neighbors cooperating to butcher hogs, cut ice, and thresh wheat. As she stated in an interview published in *The Virginian-Pilot* after she received her doctorate at Johnson and Wales University, Norfolk campus in 1996, at the spry age of eighty, "It's because of my background. We had to raise ourselves and take care of each other. And then we had to move on."[5] (See page 3 in the color insert for a portrait of Dr. Lewis.)

This illustration embodies one of my favorite Africulture principles, cooperative economics. It makes perfect sense that Dr. Edna Lewis, the godmother of the farm-to-table movement, would have her roots in this kind of soil, where there are tight connections and collaborations among food and the land and the people. For me, reading the books *Black Labor, White Wealth* and *PowerNomics* by Dr. Claud Anderson, and seeing a poster of the dollar rotation in the African American community at the Million Man March in Washington, DC, in 1995, reinforced this principle at an impressionable age. Dollars, purchases, invoices, and repeat orders water the roots of businesses, and prior to integration, dollars kept in rotation maintained a forest of African American and Afro-Caribbean businesses across the length and breadth of this nation.

Landownership is the basis for many other principles: stewardship, self-sufficiency, self-governance—all the responsibilities of caring for the land and community and descendants. Like the xylem and phloem, landownership keeps a proper tension between two opposing forces: self-sufficiency versus dependency.

Carter Farms is located on land that has been in my family for over a century. On November 5, 1910, my great-great-grandparents on my father's side, Mr. Jefferson Davis Shirley and Mrs. Catherine Walker Shirley, purchased

150 acres in Orange County for $722.05, from Mrs. Caroline D. Goodwin, who had acquired the land through being widowed.

Five years later, in April 1915, my great-grandfather Mr. John Lewis Carter purchased another twenty-five-acre property about five miles away, on what is now Shirley Road. He ended up marrying one of the Shirley daughters, my great-grandmother, Mrs. Mattie Shirley Carter Pryor. Their children—my grandfather and his siblings—were born at what the family refers to as The Old Place.

My paternal family names are on these local roads—Shirley Road and Carters Lane—because there were family bases there, now in the family for over 100 years. Various members of the families owned most of the property along those roads, many times buying and selling with each other. They were making sure they were able to provide a legacy for their children, which sometimes included financing each other's purchases. During that time period, it was sometimes illegal for white landowners to sell to Black people in some places, and a social taboo in many more places. Racially restrictive housing and land covenants were legal clauses placed in property deeds that prevented sales, rentals, or leases to a specific race or ethnicity, and these types of covenants didn't become illegal until 1968. For Black families, helping relatives finance purchases of land was a way to provide access for members of the community who couldn't buy land any other way.

Whereas Mrs. C.D.Goodwin, the Ex of the late John W.Goodwin, became purchaser of a certain tract of land August 10th, 1907, sold by John G.Williams, trustee, for certain indebtedness, the same tract was thenein the possession of B.S.Cooper and on which he then resided, referense is hereby made to the deed of the said John G.Williams, trustee to the said Mrs C.D. Goodwin, for further information; and whereas the said Mrs.C.D.Goodwin, did agree by contrast with Robert/Cooper, wife of B.S.Cooper, to remain on the place for 2 years and should have also the privilege of selling the place, by paying to the said Caroline D.Goodwin, the purchase money $715.00 and accruing interest until paid, and whereas the said Roberta Cooper did sell to Jeff Shirley, Nov.5th.1910 the said tract of land, Now Therefore This Dees made and entered into this the 21st day of January, 1913, between Caroline D.Goodwin, of the first part and Jeff Sherley and Catherine Shirley, his wife, of the second part,
Witnesseth: That for the sum of $722.05 and interest in full to date, the receipt of which is hereby acknowledged, the said Caroline D.Goodwin grants and convey with Special Warranty unto Jeff Shirley and Chatherine Shirley in equal parts all of that tract or parcel of land containing 150 acres, be the same more or less, lying and being in Orange Co.State of Va. and on which the said Shirley now resides, to have and to hold.
As witness the following signature and seal this 23 day of January, 1913.

C.D.Goodwin. (Seal)

The deed to the property my great-great-grandparents bought in 1910.

The area around Carter Farms became one of the thousands of epicenters of Black landownership and cooperative economics in the late nineteenth century and early twentieth century. People here not only helped each other become landowners, they bartered with each other and did business together. Dr. Charles West owned 120 acres next door to our farm—he was the first Black man to be a quarterback in the Rose Bowl, during the 1920s, and was also a football coach at Howard University. He became a medical doctor in the 1930s. Dr. West provided medical services throughout the community. Another Black family, the Terrells, owned 500 acres next to our land on the west side of Carters Lane.

From stop sign to stop sign, the land was owned by African Americans—a community with about 800 acres of land altogether. That's a strong set of roots, and a good little amount of financial security. When the Depression happened, they were insulated because they had all their needs in place in terms of food and shelter. There's a stream for a water source, and several hand-dug wells. There was meat—livestock such as chickens, pigs, and cows, and wild game such as pheasants, rabbits, wild turkeys, squirrels, possums, and even snakes if you wanted to push the limits—and they were walking distance from a couple of ponds to fish in.

After Gran married my great-grandfather, they began raising their own family. Uncle Joe, who transitioned in January 2023, was my grandfather's youngest brother. He's the only one of his siblings who's really given me stories about his father and how they arrived at the land. He was born in 1936 at The Old Place. On April 12, 1945, the same day that FDR died, the family began their Sankofa return back to Gran's parents' land, where they lived after they had paid off the debt on that property. (I tell the story of how the debt was paid in chapter 4.)

Uncle Joe told me that he, at the ripe old age of nine, and Aunt Faye were instructed to bring four cows down to their new farm, and they had to go the long way, off the roads, so that they wouldn't be perceived as having stolen the cows. As Black youth, then as now, they suffered the same social prejudice of potentially being guilty until proven innocent simply for going about their daily life.

At the time, the farm had a lot more trees than it does now. Uncle Joe told me his brother Uncle John, fifteen or sixteen at the time, began clearing the land. To somebody of my generation, fifteen or sixteen years old seems young to do that type of work, but during that time the mindset was just, can you do the work? At that age, you're strong, you've got an ax, and you're obedient, so if mama said clear this, you cleared it. When his older brothers, including my Grandpa Warren Carter Sr., got back from the war, they also helped to clear trees, create pastures, and set up the homestead for the next three generations.

ROOTED IN RESPECT

The women in my family provided very strong examples of self-determination, vision, family building, and community-based independence. I didn't recognize or understand this when I was younger, but while researching my family, I've come to realize that my ancestors were making a point to include women in the ownership of land. Every deed that bears my great-great-grandfather's name, his wife's name is right beside his. They were always very intentional about that. Ownership and family were approached as a team. You wouldn't marry someone you didn't consider to be a good team player or worthy of co-ownership of your assets. Marriage for Black families provided rights and security for both the husband and the wife. "Til death do us part" was a literal instruction. Divorce and separation were nonexistent in my family legacy in the early twentieth century.

As for the men, they set a high standard. It wasn't until I went to college that I met people whose fathers weren't part of their lives. Growing up, I knew a few people with divorced parents, but most of these fathers were actively present in their children's lives. Almost all of my peers were from married households. My introduction to manhood was through these men. The emphasis was on responsibility and accountability for your actions, and that made a tremendous impact upon how I approached life. It was so prevalent, I didn't know any other way of being a man. These men had a joviality, but also a seriousness when they needed it, and a readiness to live up to standards. If you ever get someone pregnant, you are going to take care of that baby. Africulture defines manhood, womanhood, and adulthood as the ability to be responsible and accountable for your actions and the actions of others in your care.

We recognize the importance of accountability and show respect to adults in the way we speak to them. Everybody in my world receives a title as a sign of respect; everybody over twenty-five or thirty is an aunt or uncle. Visitors on my farm will hear me mention Uncle Roland, Uncle Gib, Uncle Carlton, Uncle Frankie, Uncle Little John, Uncle Eddie, Mr. Bates, Deacon Howard, Uncle Clif, Uncle Glenn, Uncle Louis, and Uncle Vincent with the same respect and honor I'd give my birth uncles. Even if an adult said to me, "Call me by my first name," I would have a tough time doing it—if I did so at all. In West Africa, too, elders are addressed respectfully. This matter of respect, of sawubona, was a requirement for me. This respect isn't just relegated to those older than me; it's quite customary to hear me refer to my sons as sir, and to greet a young lady as ma'am. Respect is a birthright that we extend until someone behaves in a manner that shows they don't deserve it.

In Black communities, it is essential to be respected, because you never know when that respect could slip away again. Sometimes my great aunts and my grandmothers would become upset when they talked about working as domestic servants and being called by their first name by their employers' children. That was always a sign of disrespect, but you couldn't do anything about it. It harkened to the days of human trafficking, when it was common practice to deny people names or any other kind of acknowledgment. In the Cape Coast and Elmina slave dungeons in Ghana, there is not one mention of any names of the hundreds of thousands, if not millions, of people who passed through or died there.

Different cultures are like different soils, and plants (or people) respond in kind. You have to adhere to the nature of what that soil is so you can grow and survive. A taproot has to grow to one depth in sandy soil and another depth in clay. The culture I grew up in had deep taproots into the land, into a heavy clay with opposition at every level, and I grew deep roots anchoring myself to these older people. Respcct, like the sap in the xylem and phloem of plants, should flow both ways for maximum health.

Within our own community, the rule is to show respect to others by calling them Mr., Mrs., Dr., Uncle, Aunt, or Cousin. It's part of the roots—that respect for your nature. For a child to call you by your first name is to not respect you as a man or a woman.

MYCELIAL NETWORK

The roots of most plants are supported in a multifaceted way by underground fungal bodies, often seen as networks called mycelia. Like plant roots, the threadlike fibers of mycelia absorb, collect, and distribute water and nutrients. These structures spread resources across forest floors, under roads and pavement, and they assist in the decomposition of nonliving or dying elements. Metaphorically, the roots of people need the support of water and mycelial networks, too. We need resources to help us stand up straight and we need relationship networks to feed us.

In Orange County, my family had been part of a Black Wall Street of small Black-owned, community-supported businesses. It wasn't only big urban areas such as Tulsa, Oklahoma, that had Black Wall Streets, thriving economic self-sustained communities. Because of segregation, Black communities across America had to rotate dollars among themselves. Members of their respective communities were limited to working as sharecroppers or domestic workers, but they could have additional income through running small businesses that were supported by other members of the community. They didn't always reveal their monetary resources in the form of buildings, but they might have money kept under the mattress or invested in land.

ROOT BOUND

If a plant crop is having trouble, we often look to the soil to figure out what the problem is. Are there pathogens in the soil? Is the pH wrong? Are there harmful nematodes, or some kind of pollution? Has the ground been overtilled, destroying the soil structure? Is it receiving enough rain?

If Black farmers are the "crop," we need to ask the same kinds of questions. What's in the environment that's making it so hard for these families to thrive as landowners and stewards? Most of us are familiar with the Japanese practice of bonsai, which literally means planted in a container. This practice is parallel to the African American experience in agriculture and, on a macro level, our general experience in America. Our roots have been bound, cut, pruned, and ended up stunted inside containers we've been forced into. Slavery, indentured servitude, sharecropping, lynchings, apartheid, colonialism, imperialism are all vessels that have limited their victims' root growth. America, and Western civilization more broadly, is a container that limits the fullness of the African genome and its genetic expression.

With any example of racism in America, there are legal factors and cultural factors. The legal barriers to Black farmers were many, and I explore those more in chapter 7. The roots of this social foundation included the three-fifths compromise written into the Constitution, the process of land theft, and the redistribution of the stolen land to white homesteaders. It includes the preferential way agricultural education and support was offered disproportionately to white farmers, the fact that farm and home financing was unequally available based on race, and that occupations often held by Black workers were left out of minimum wage and social safety net legislation and policies.

In reality, you can't separate the legal from the cultural. One is rooted in the other. In the same era during which Supreme Court Chief Justice Roger Taney wrote, in the Dred Scott decision, that Black people "had no rights which the white man was bound to respect,"[6] some high-powered Americans formed a secret society—so secret, that you may never have heard of it—that sought to create a slavery empire. The society was called the Knights of the Golden Circle (KGC). Founded in 1854 by George W. L. Bickley, with notable members like John Wilkes Booth (killed in my home county in Virginia after he murdered President Abraham Lincoln), Freemason icon Albert Pike, and Vice President John Breckinridge, the empire the KGC envisioned would have included the territories of the southern and western United States, Mexico, Jamaica, Cuba, and most of the Caribbean and Central America. The capital would have been in Havana, Cuba, and the KGC sought to control the trade of sugar, cotton, tobacco, rice, and any other commodity crop of the period, with a multinational system of enslavement as its fuel, glue, and foundation.

During the Civil War, the KGC rebranded as the Order of American Knights and then to the Order of the Sons of Liberty, ultimately merging their efforts with that of the Confederate states. While they never accomplished their imperial goals, they became one of the model organizations for America's first terrorist organization (although it has never officially been labeled as such), the Ku Klux Klan.

BLIGHT

One of the reasons I started Africulture was to challenge the perception of people of African descent as related to agriculture. The invisible depletion of Black farmers over generations has a lot to do with the atmosphere in America, the blight in the soil of our society, that grossly affects and infects our roots. There are unwritten or taken-for-granted elements in our culture that for generations have been telling young Black people to do something other than farming.

In the early twentieth century, when my ancestors were farming here in Orange County, they were thriving in many ways, but at the same time they had to be very aware of their profile in the wider community. Because of the cooperative economic principle, money rotated locally and sustainably, with no support from banks or the broader white community. When someone started a business or did something for extra income, selling cakes, pies, or chickens, or doing home repairs, there was an instant market. The community supported new ventures not just financially, but also with know-how. As people went through the process of growing produce, raising livestock, sewing, bartering, trading, selling, and investing in themselves, capital was quietly accumulating in their communities. Spending time in Africa, throughout Ghana, from Wenchi to Sogakope to Cape Coast, I witnessed similar community development, involvement, and support.

My great-grandfather Mr. Guilford White and other family members.

African-American farmers in the early twentieth century tried to be as low-key as possible. You supported your family, but you did not put on the air that you were too successful. A car was a sign of success, and it would draw extra attention. "What are y'all doing to be able to afford a car?" If you had a car, it had to be used and dilapidated—a 1909 Model T and not a 1919. You had to show you were scraping pennies together to make this happen. It was what I term the Booker T. Washington model—be essential, but not flamboyant with your essential nature. As a result, Black farmers were silently barred from showing business growth, and any materialistic signs of success. The inherent success could invite night riders, not the space-age car from the 1980s television show, but the Ku Klux Klan or other terrorist cells. They would come to damage your property or attempt to run you and your family out of the community, or worse. Men and women were forced to diplomatically protect their peace and property by portraying themselves as staying within their perceived rightful place in society, which was some form of poverty and struggle.

While a thriving local economy ensured these farmers' survival, the flip side was that it was near-impossible to do business outside the community.

ROOTED IN TERROR

Most people know something about the history of lynching in America, the influence of the Ku Klux Klan, the outright oppression and segregation during the Jim Crow era. My family has our stories of threats and worry from that time. For one, when my Uncle Joe was born, apparently he appeared to be lighter in complexion than a Black baby should be. He told me that he later learned that people in the community thought he was too white to be Gran's baby, implying that she had had a relationship outside of her marriage with a white man. They threatened to come out and whip her for having a white baby. Uncle Joe was the youngest child in the family, and his father gave instructions to his older sons: "If the Klan comes to visit us, I'm going to go out and talk to them. If I go down, you shoot all the SOBs."

There are other stories I haven't been able to corroborate. In the 1940s or '50s, a white young lady apparently had a teenage, seemingly innocent crush on one of my great-grandmother Gran's siblings or nephews. My great aunt Lucille remembers that her father saw the nephew and the girl coming out of a barn and said, "We'll go get that boy and put him on a train to Philadelphia."

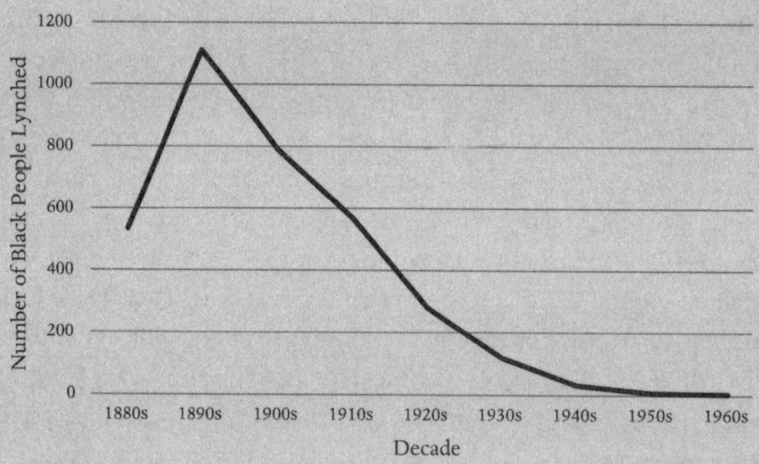

Lynchings over time. *Statistics provided by the Tuskegee University Archives.*

Apparently, though, my great aunt's father wasn't the only person who saw them coming out of the barn. Word got back to the girl's father, and he planned to make an uncourteous call to this innocent young man. The family moved fast to get the young man to the train station that evening, and he never came back to Orange again. He was about sixteen or seventeen at the time.

This kind of terror was felt by so many Black families in those years, and many were traumatized by actual violence. That affects people's roots and connection to a place profoundly. Feeling safe and secure is a prerequisite to thriving; when you don't have that, you're in survival mode. And it was that threat of violence, along with the hope of economic opportunity, that drove a lot of Black farmers off the land to Northern cities during the Great Migration. This wasn't the only factor but one of the major factors indeed.

At Carters Lane, my family always had cows, chickens, pigs, and standard vegetable crops for their own use. But they didn't grow commodity crops like corn because they didn't have a market to sell them. Discrimination was and is a core component of the American soil that chronically infects the roots, hindering the exudates and the absorption of nutrients that make the community grow. Even if a white farmer would buy your produce or grain, you couldn't guarantee he would be fair with you. Some were fair and upright, and many more were not. This problem is still occurring for some Black cattle ranchers when they send cattle for slaughter. To avoid being shortchanged by the processing facility, a Black rancher may need to send their cattle with a trusted white farmer, who has a larger operation, to ensure they will get a spot for their animals to be processed and then to ensure they received fair pricing. Some of the challenges Black farmers have faced for too many generations have evolved into nuances, and are simply considered the cost of doing business.

Other reasons for the decline of Black farmers were a little more subtle. I think about my own experience growing up—the messages I got about my future. As I've said, I grew up steeped in agriculture, with my father and so

Table 3.1. Decline in Black Farmers 1900 to Present

Year	Number of Black Farmers in the US	Number of Black Farmers in Virginia
1900	746,717	26,566
1920	925,708	47,786
1925	*	50,147
1940	681,790	35,062
1959	272,541	15,629
1978	37,351	3,978
1997	18,451	1,241
2017	35,470	1,335
2022	32,653	1,079

* No data available.

Data compiled from the USDA National Agricultural Statistics Service archives, https://www.nass.usda.gov/AgCensus/archive/files.

many other role models being ag teachers, but agriculture was not presented to me as a viable option for a career. I remember riding through Cumberland, Virginia, on the way to my aunt's house. The poultry houses along that route reeked with fumes that singed my eyes and nose hairs, and possibly my ear hairs. It was like being inside an ammonia bottle, and I'd have to hold my breath and endure this noxious odor. Those "farms" did not come across as inviting places for a nine-year-old to visit, let alone think of being employed at. I was never told, "This is a business you should look into." I would have possibly been interested had I been informed of the amount of revenue generated from one of those poultry houses. But what I would hear from farmers I knew was "I don't make no money." To a child, that's the most unattractive thing you could say. I was taught to get a good government job—that was the safest path for young Black people with college degrees. Many people in my community in my childhood were teachers and FFA advisors and 4-H advisors, but at that time, most of them never made agriculture sound like an attractive way to make a living.

Numerous Black male middle and high school students would veer off into sports or other extracurricular activities as they progressed through school. You had to be dedicated to the craft of farming or see a future for yourself in the industry to stay.

As I got older the racial bias and cultural differences became much more glaring. There was country music and rock music blaring from vehicles in the parking lots of FFA events I attended, instead of hip-hop or go-go or whatever I was listening to in the early '90s. Nothing was less inviting than being at a remote location in central Virginia in the late afternoon, about to enter the forest for a tree identification contest, and looking over at Confederate belt buckles and hats leading the way. Even though I never experienced any form of discrimination or racism while participating in FFA activities, I was not interested in playing racial roulette at dusk to win an FFA contest.

Once I got to college, there was a different form of pressure not to be a farmer—even though I attended a historically Black 1890 university. Most agriculture universities, specifically HBCUs, didn't promote farming as a full-time occupation at the time; they promoted agriculture industry jobs, as I discovered when I was attending school in the late 1990s. We were taught how to get a job, create a career in agriculture, and encouraged to seek employment at the USDA or in the agriculture industry. At the time, I wasn't encouraged to be a farmer or landowner. I wasn't discouraged either, but the emphasis was placed much more on the industry side. And rightfully so, as the average annual revenue for many Black farming operations is under $30,000 in most states.[7]

This is not to disparage HBCUs. They're doing what they feel is best for their students to acquire careers for security and success, and they are seeking to pave the way for future farmers to walk down the paths laid by these trailblazers. The 1980s and 1990s were a time when decades of federal government discrimination against Black farmers were coming out into the open. Our roots in the agriculture industry had been stunted, cut, and poisoned, and it was necessary to bolster the root stock—the agriculture students at HBCUs—to better handle this potentially toxic environment.

It's different at predominantly white institutions, where I've heard from students that "I'm going to school for my farm." Many of them are already rooted in their own family farms. I recall visiting a poultry processing factory while I was in college. As we were traveling to the factory there was a putrid, foul smell in the air. A white classmate of mine said so clearly, "It smells like money." To him it was production; it was revenue, an understanding that something as common and foul as fecal matter had a value in agriculture. I had not made that connection before. The fact that a nineteen- or twenty-year-old knows that a foul smell can produce revenue says something about his mindset toward farming as a business.

I didn't want to work for the agricultural industry, but I also didn't think about being a farmer at all. I followed my interest in fashion design and lived in a big city and then I moved overseas. It would be several years before I got back in touch with my roots in small farming, and even longer before I regained my connection to the land in Virginia, where my family's been anchored for so long.

THE ROOTS OF HEALTH

As younger generations of Black farming families have moved away from the farm, cutting off their roots in the land, the cooperative economy of Black agrarian communities has withered. Black farmers lost their Black customers, who had been their main markets. The Great Migration affected not only people's financial picture, but their health. My grandmother Mrs. Lucille Carter ate from a little garden in the back of her house, employing limited tillage. Her family pulled sweet potatoes right out of the ground and picked ripe tomatoes off their own vines. They grew string beans, snap peas, and sweet corn, and they ate a lot of wild foods, too: wild blackberries, blueberries, poke sallet, and plantain leaves. My great-grandmother used to make dandelion wine and they used black walnut hulls to make a tea that could purge the body of parasites. There were greens on the table for every meal except breakfast.

My grandmother was also a huge believer in canning. If you were at Grandma's house, you took part in the harvest. When we visited her we had to pick string beans and then string them and snap them. Or we'd be cutting apples, pears, or tomatoes. I didn't appreciate it at the time, but she didn't buy canned spaghetti sauces—she made it from her jars of canned tomatoes. In the years before I was born, my family maintained a much lower level of meat consumption, with almost all of it grass fed and local. Meat was reserved generally for Sundays, and it wasn't as plentiful as it was when I was a child. The segregation policies of the time kept them from being able to access the same groceries that their white counterparts could. It was one of those things that in my opinion helped sustain their longevity—at least for the females. In certain respects, segregation was beneficial for Black economic development and health. Eating out at restaurants wasn't a frequent event, so that money would be saved or spent with someone local. Limited access to supermarkets also limited their access to processed foods that contained additives and preservatives. The civil rights movement initially was not about action against segregation as much as undoing the premise of separate but equal. Equity, fairness, and being treated equally was what the movement initially desired, not necessarily the right to spend money outside of their community. The movement did not morph in that direction until its later years.

It's ironic that one of the results of the Civil Rights Movement and the end of Jim Crow was that it flipped the script for how a lot of Black people eat, at least in the rural South. Now that African Americans could partake in the processed foods for sale at grocery stores, those foods became more prevalent in their diets. We started to eat like white people. Because we hadn't been allowed to go in grocery stores previously, shopping in an integrated store became a sign of success for some. There were also marketing and psychology ploys related to convenience of food preparation, especially as the workplace allowed for more minority and women employment. Feeding your family was still at the center of life for many households, but the more convenient it was to provide processed foods that could be cooked and prepared in minutes instead of hours gained greater value.

As many Black families changed their shopping habits, it eroded support for Black farmers. These changing nutrition habits, changes to the very soil that nourishes the roots of Black people, are tied to big trends in agriculture—how farms have become big businesses rather than the means for families to be connected to their land and subsistence. The "money" that the white student had smelled on that college trip was profitable indeed, but it wasn't a model for good health. Big corporate agriculture is much more focused on profit than on people, nutrition, and the environment, but it's the standard for agriculture that the industry stands upon. Even when we're talking about environmental support or needs, the first thing that pops up is: How is our profit affected? The good of the planet is not put in first place; profit margin and stakeholder value are the priority for most of the industry.

This is all reflected in what people eat. Sixty-seven percent of our children's food supply is ultra-processed foods. The base is some combination of corn, soy, high fructose corn syrup, cottonseed oil, sugar beets, and maybe some oats or canola oil. It is foods like these that are going to drive us further into the need for health assistance. You cannot make ultra-processed lunch meat or cheese in your kitchen or even your high school science lab or a commercial kitchen, but these products are now staples of our diet. This foundation of processed foods is the "soil" that's affecting our roots and the biological pathways that supply—or fail to supply—our bodies with nutrients. Packaged foods from the grocery store don't come from the soil where we live. Most of them are grown and processed far away—so when we eat them, we become uprooted from our homes.

In America now, the offerings at the grocery store dictate what we think of as food and what we eat. In many other countries, that's not so. It's the food on sale at a farmers market or open air markets that determines what you eat. In Ghana, the only processed food I ate was spaghetti noodles and tomato paste (that contained only two or three ingredients), and everything else was whole foods. When we went to the market and we bought peanuts, we would take them to the mill and have them ground into peanut butter. If we got soybeans, we would use the blender to turn those beans into soy milk or tofu. It was great food, and I knew exactly what I was eating. I could identify the sea that the salt came from, and possibly the farmer who worked at the salt farm that harvested the salt.

America's a unique place where we've created a food industry that's the center of our existence and yet not a source of healthy sustenance. We now have to consciously fight to identify what is and isn't a food and what is and is not healthy, versus what is affordable and convenient.

Historically, my family's had issues with prostate cancer, and also a history of diabetes. My Gran had diabetes, or sugar, as they called it back then. She had gangrene set in and had to have a toe amputated. Then came the amputation of all the toes, then the foot, then to the knee, then the other leg—the same progression.

In September 2003, I went to see my other great-grandmother, Grandma Lizzie. I heard her say to someone, "I want Michael J. to cut that mess off his head." She was referring to the seven years of locks I had growing from my head at the time. I was battling within myself to make sure that I was being me, but I couldn't let one of the last requests of my beloved great-grandmother not be fulfilled. She had a slew of children and grandchildren, and then a slew more of great-grandchildren. It was a lot of people, and the fact that she would take note of my hair meant something to me. So I had

Grandma Lizzie and Grandpa Gil.

my hair cut, and as it happened, the next week she went into the hospital. I went to see her one last time, the week before she transitioned at age ninety-three. I said her name and pointed to my head, and she smiled. It was a look of approval but also, "Thank you."

After she died, I remember so vividly watching the prep for the funeral and people cooking—ladies with huge ankles and swollen legs, baking ham, making mac and cheese. The smell of pork invaded my nose like America had just invaded Afghanistan. I was twenty-five and I thought, "We're going to continue this death march at the funeral." My paternal grandfather, who shared a birthday with Gran, March 5, had the same health scenario and was losing his battle with diabetes as well, the gangrene in the exact same progression of amputation, from the toes to the foot to the leg, to the other toes, foot, and leg. I didn't want that at all. My father was having health issues, too, and a few years later he was diagnosed with prostate cancer; a few of his uncles and cousins have had it and his grandfather died from it. I didn't want this as a future. I postulated that if I kept eating the way my family ate, I would have a similar fate to my grandfather and great-grandmother.

The food system I would like to see is less efficient, but healthier—getting back to small-scale farms on large-scale acres, with a focus on nutrition and lifestyle. The regenerative practices of agriculture would be for the soil, people, and animals. More people would be working manually on the farm, eliminating the use of certain technologies that have destroyed soil structure and transforming the monocrop system of food production into a sustainable biologically integrated system. As far as what we eat, I would like to see 50 to 60 percent of our food in stores naturally grown from a farm and slightly modified for value-added products, something you can do in your home kitchen.

The trees at Carters Lane—any woods—are the pillars of the environment, deep-rooted anchors for everything else that lives there. When an elder dies it's like a pillar, a mighty tree, that's been knocked down. It's so easy to lose touch with our roots in different ways. We will literally chop out a root if it sticks out of the ground too much. If it disrupts the sidewalk, it's coming out.

But respect for the roots is respect for history. Principles and values are like roots keeping everything else intact, keeping culture, civility, and basic respect from eroding. If you don't have a relationship with the roots, you may treat your own health or future like a tree that's in the wrong place, about to get cut down. The roots of Africulture are not just for the sake of being sticks in the sand, they're designed to keep the substrate, the other aspects of our culture in place. And when you lose those roots, lose those values and principles, the structure and the integrity of that culture dissipates and becomes useless.

That so many African geniuses germinated and grew despite the plethora of challenges facing them is an amazing accomplishment that speaks to the turgor of pressured roots and the potential of even the most restrictive environment. It also speaks to the power of the mycelial networks that support root systems. The roots of American agriculture were intentionally bound and manipulated to stunt the brilliance of its African and Indigenous population, but their roots could not be suppressed, and their growth and subsequent fruit continues to astonish and amaze.

It's often mentioned that Black farmers in 1920 owned over 15 million acres of farmland. They owned this land outright, without any financing from mainstream banks and with limited or no support from the USDA, as segregation and "separate but equal" was the rule of the day. The community had a determined taproot, and many adventitious roots that sustained their survival and growth in the late nineteenth and early twentieth centuries, despite the many pathogens, stones, and hardpans in the soil. And these obstacles are still with us, as later chapters in the book describe.

CHAPTER 4

COMPOST

WHAT ARE YOUR INPUTS?

When your mind is focused and made up you don't make excuses.
You try. If you fail you try again.
Minister Louis Farrakhan

Composting transforms energy loss from one organism into energy gain for another, through the metabolic activities of quadrillions to quintillions of microorganisms—a Million Microorganisms March. The microorganisms take the scraps, the manure, the fallen, the dead, the damaged, and recycle it into something that can support the growth of other organisms. There is no life without death, and this deathly sacrifice is popularly referred to as the circle of life.

"Limited inputs" is one of the principles of Africulture that speaks to the importance of proper recycling. If you don't reuse the materials that your farm provides for nutrients, you'll have to bring in artificial inputs such as fertilizers. Limiting the outside inputs means making use of what you have in the most productive way you can. The flip side is that the proper disposal of whatever waste ends up on your hands (setbacks, loss, sickness, death, and hatred) has to happen. If you don't deal with the waste properly, it may attract unhealthy vermin, flies, roaches, and mice. In human terms, these are conditions such as low self-esteem, stress, preventable diseases, depression, and early death—among others. Composting our traumatic experiences is a way to transform them.

On the farm, compost generally involves two components: carbon-based inputs and nitrogen-based inputs (the former are commonly referred to as the browns, such as wood chips and dead leaves; the latter are the greens, such as grass clippings). A balance of more brown inputs to green inputs is needed for the proper breakdown of the material, which can be an aerobic or an anaerobic process. Aerobic decomposition is carried out by microorganisms that use and require oxygen. These microorganisms release oxygen as they respirate, which allows other microorganisms to thrive, making the transition process more efficient, while assisting in eliminating unpleasant odors and stenches. Respiration releases energy, causing the compost to heat up as the microorganisms increase in number, leading to more and faster breakdown of materials into useful organic matter. Vermicomposting, traditional compost piles, and most of nature utilize aerobic digestion.

Anaerobic is the opposite process: It is the mode of decomposition by microorganisms that don't require oxygen to transition the waste matter into something useful. This composting process is slower, and an anaerobic pile sometimes becomes stagnant and stops breaking down altogether. Stagnation causes odors to persist and temperatures to cool. Breakdown of materials happen slowly and inefficiently. There are methods of composting, such as bokashi, that deliberately rely on anaerobic decomposition; and anaerobic waste digesters extract methane gas from decaying waste and fecal matter to utilize as an energy source.

Composting is also a metaphor for something melanated people from around the world have continued to do: taking the worst of modern existence and making it usable and viable. It's a social alchemy. And we have utilized both aerobic and anaerobic methods. Some do best by breathing, inhaling, and exhaling to allow oxygen to fuel us. Others hold in the breath—persisting despite the lack of oxygen, the lack of opportunity, dealing with odors and stenches and taking a longer road to get to the desired product. Whichever process is chosen or inherited, the urge is to deal with the manure of microaggressions, mismanagement and malice, the dirt of discrimination, the accusations of inferiority and worthlessness, the organic insults and misrepresentations—and turn it all into something that fuels growth, progress, and success. I've witnessed and heard numerous stories of our people turning manure into gold, a feat even more phenomenal than changing lead into a precious metal.

For plants and people, compost represents a layer of life that we have to get through, taking what's helpful and pushing past the harmful (vile and ignorant attitudes, behavior, policy, and/or actions) before we can grow toward the light. When compost gets put into the soil it seeps down into the roots. A plant's roots determine how much of this compost it can absorb. Likewise, a person's roots determine how much they can grow through and still be in good health. The deeper, longer, fuller, wider the root system, the more nutrients they can take up. Compost is adversity, but it's also nutrition. Like compost, the adversity that is faced can be vulgar in its appearance and fragrance, but over time, if handled properly, some level of good or learning can be extracted from it. The experience of dealing with challenges can make you richer and more complex—if you can overcome it.

As discussed in chapter 3, Freetown and many other rural and urban Black Wall Streets across America became strong, interdependent, self-reliant communities in response to segregation, social stigma, and oppression. They were marked by cooperative economics, a social standard, food production, and the spirit of communalism. These communities flourished because they pushed back against the conditions of adversity that surrounded them. Their practices of communal and social alchemy turned the garbage that was given to them into social jewels such as values, community, standards, protection, and hope.

These places were made up of families—if not blood family, then community family—each one dealing with its own problems. For my family, like many others in Black farming communities, landownership was an important part of being able to transform societal compost into fertile, productive lives. But that didn't mean we were guaranteed the ability to hang on to our land. The property where Carter Farms sits today, land that my great-great-grandparents purchased in 1910, came close to being lost by our family just a couple of decades after they'd made their home here.

No one living knows the exact story, but near the end of the 1920s the land was used as collateral on a $636.40 debt. I've heard talk that maybe it was to pay for a wedding, but whatever the reason, soon afterward, the Black Thursday stock market crash of 1929 happened and the Great Depression began. The individuals who had borrowed the money couldn't pay back the debt. My great-great-aunts and great-great-uncles were brought to court over this matter. If we had lost the farm, I wouldn't be writing this book, and maybe I wouldn't even be here in this realm of existence.

Ten siblings were in that generation, and Gran was one of the youngest. I didn't know this when I was growing up, but my great-grandmother ended up being the key person who saved that land from being lost to the family.

Mrs. Mattie Shirley Carter Pryor, or Gran.

I never heard this story about Gran from the elders who raised me. There wasn't exactly a code of silence when I was a child, but there was definitely a code of age-appropriate answers. Some topics weren't shared with children, or adults who were potentially seen as children. And the historic relevancy wasn't seen as significant enough to share or to use as a teaching tool. The value of our family history sometimes got intermingled with the refuse and garbage received socially and just became part of the finished product of compost. In other words, my elders just viewed it as "what happened, happened," and didn't regard it as a story to make a fuss about. Sawubona Gran.

It wasn't until 2019, two years after I had moved back to Virginia from Ghana, that I pieced together the story of the debt on the land. I inquired with members of my family about the land's history. My curiosity started to bring out stories and information, and this fueled me to visit local courthouses and piece together more of our history with official court records. (I encourage anyone looking for information about their family, especially related to landownership, to visit their local courthouses and dive into the records.)

To my surprise I was able to find deeds for the property from the late 1880s, before my family owned it, all the way up to the late 1970s. I found verification from court records that my family had owned this property since November 5, 1910. A narrative started to take shape: We've got the land here, but then something happened in 1929. And in 1933 we were in probate court over this debt. Sawubona Grandpa and Grandma Shirley.

How did we avoid the loss of the farm? By that time, my great-great-grandparents, who had originally purchased the property, were gone. Their ten children were now the property owners, but most had moved out of state. Gran was not listed on the 1920s deeds and court documents. She was

My grandfather, Mr. Warren Harding Carter Sr.

My great-uncle Mr. Roy Coolidge Carter.

thirty-four at the time, with several children, and still lived nearby on Shirley Road. In 1937, she took the place of her siblings as the lead name on that deed. The other heirs didn't want the land, and they were willing to let it go. But Gran wasn't. She knew the county clerk, Mr. DeJarnette, and he gave her the right of first refusal and helped make it possible for her to pay off that debt over time. She made a commitment to ensure that the land stayed in the family.

Soon after that, World War II broke out. Her three oldest sons, Warren Sr., Alphonso, and Roy, were drafted. And for the two or three years they were in the war, they all sent back the $22.50 per month they were paid. Gran's daughters worked as domestic workers in local residences and also contributed their pay to keeping the land in the family.

Does cannon fodder serve as compost? As the draft took hold of the country, it touched most families that included young draft-aged men. Black draftees had to accept the social garbage that they were citizens enough to sacrifice their lives for the war effort, but not citizens enough to have full rights of equal treatment. The three-fifths compromise hadn't yet been broken down in the compost pile of America. Uncle Alphonso was in the infantry as a driver in Wales when he first entered the war. While I was writing this manuscript, I heard from an English documentary team who were researching a film on black GIs in Britain during World War II. They shared a story no one in the family was aware of. In 1944, Uncle Alphonso provided a testimony under oath to US military police about the murder of John Lee Hendricks, one of the Black GIs in his unit, in Cardiff, the capital city of Wales. I understand that the incident stemmed from the local British population having the audacity to treat the Black soldiers as comrades, with dignity and respect and, by allowing these soldiers into a neighborhood bar, choosing not to apply the deadly virus of racism and discrimination under the guise of separate but equal to their public life. My uncle and four fellow Black soldiers from the 1938th Quartermaster Truck Company were leaving the bar after a dispute with other US soldiers when they found themselves confronted by the American Military Police, who fired on their truck as they were driving away. More than three dozen white soldiers, sailors, and MPs then forced the Black soldiers out of their vehicle, threatening to kill them all. Mr. Hendricks, wounded in the shooting, spent approximately thirty minutes to an hour bleeding out in the cab of the truck while my uncle and his fellow private first classes were interrogated publicly and arrested for the crime they had been guilty of all their lives, the felony of Blackness.

"We could do nothing towards getting John L. to a doctor," Uncle Alphonso testified. "At this time they had guns pointed at us and I had heard them cock same [some]." Everyone except the injured man—who, without treatment, soon died—was taken to jail. "Before taking us away and while we remained standing there with our hands still over our heads, they made bad remarks: 'We should kill all those black sons-of-bitches.' Some said 'black mother fuckers' and others said similar remarks. There was nothing that we could do . . ." Sawubona Mr. John Lee Hendricks.

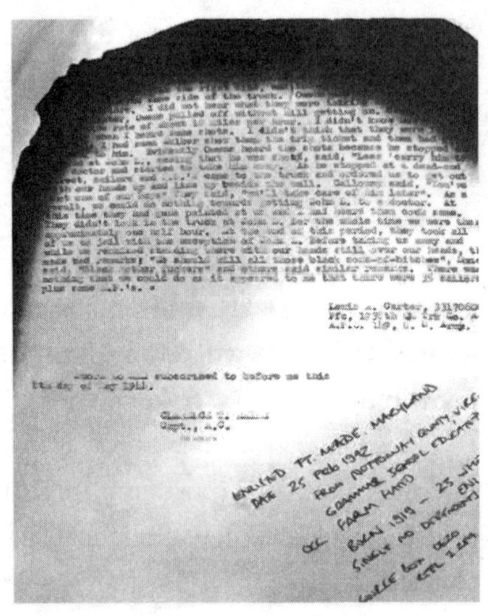

The typed document of my great-uncle Lewis Alphonso Carter's testimony, singed all around its top edge by fire.

My Uncle Alphonso's testimony, a piece of history that came very close to being lost in a fire at a military records facility in 1973, is one undeniable example of the hate that Black soldiers faced from their own fellow Americans while they were serving their country far from home. This was not "friendly fire," but very much enemy fire, even though they all wore the same uniform.

This experience and many others are examples of anaerobic composting. No oxygen was given to this experience, no light of day, no mention to his family, although it may have been shared with some of his war brothers. It was a horror that Uncle Alphonso literally took to his grave. It was so putrid it was left to rot, only to be discovered fifty-five years after his death. Now it seems it's the right time for that story and experience to be digested and broken down within the family.

Grandpa Warren Sr. served in Europe in the Quartermaster Corp Graves Registration Service, the unit that managed the administration and burial of soldiers, one of the most dreaded assignments in warfare. The majority of his experience in this war was working with dead bodies or mortally wounded men, mainly in hand digging graves. I imagine that he would have constantly worried about finding himself burying one of his own brothers, or any of several other friends, cousins, or neighbors from Carters Lane stationed in Europe, as well.

I never talked to my grandfather about his experiences. I didn't even know he had served in World War II until he was close to his death, when his daughter asked about making sure a flag would be put on his coffin to show that he was a veteran. No pictures of him in uniform were displayed in Gran's house or my grandmother's house. As a boy, I knew my grandfather as a heavy drinker and snuff chewer; every time I saw him, he had on his overworked overalls with one buckle hanging off, spit can in one hand. It wasn't until I was older that I began to imagine and empathize with his earlier life. He was drafted to serve a nation that gave him no rights; he couldn't vote, eat at certain restaurants, or use certain water fountains. He truly couldn't be caught in some areas after sundown, because like many in that time, his skin was his sin. I reflected on the lack of empathy he received and the perceptions of him from others, not knowing his story, and the bull crap he endured. He embodied the words and sentiments of Marvin Gaye's musical classics, "What's Going On" and "Trouble Man." Sawubona Uncle

My great-aunt Mrs. Anna Carter Scott and Mr. Lee Scott Jr. (a WWII veteran and Tec 4 in the US Army) on their wedding day.

Alphonso, Grandpa Warren, Uncle Roy, Uncle Ernest, Uncle Jeff, Uncle Lee, Uncle Landon, Deacon Floyd Bates, Deacon A.B. Howard, Uncle Woodrow, Uncle Kenny, Uncle John, Uncle Al, Uncle Joe, and Uncle Frankie.

My great-aunts, working as domestics during that time to help their mother pay off the debt, faced similar discrimination and restriction. They were young, in their late teens or early twenties. Aunt Anna Scott would vehemently bristle up every time she heard the name of the woman she worked for. Visceral anger and disdain would cover her face, followed by a few choice words that couldn't be repeated on 1980s television shows or around my house. During this time period, though it was never mentioned, I can imagine there was a certain level of sexual abuse or assault, at least verbally, from those families they worked for. As a Black domestic worker in a white family's house, who's going to help you? You are automatically disqualified from being treated with respect. Being at the mercy of white people, who had the dominance and power, were the reality of these situations for Black women and girls. The fear of retribution was and is real. And there were the microaggressions, such as the false praise—"How come other Black people [generally not the term used] can't be like you?" But my great aunts, like many Black teens or preteens in this era, needed to take these jobs to add income to the family household. Most were encouraged to keep their head down and wear their mask.

I think of those examples of Gran's daughters, my Aunt Marjorie and her sisters: They utilized those funky elements to help grow. And their growth helped to plant other seeds. Aunt Marjorie helped all of her sisters and many of her community friends, nieces, and nephews make a transition into living and working in Washington, DC. She allowed them to stay at her home and get on their feet. These young aspiring adults grew in the compost that was created by my aunts' sacrifices in the 1930s and '40s and became fuel and compost for others.

My great-aunts, Mrs. Bettye Jones-Isom and Mrs. Marjorie Lightfoot, in the mid-1940s.

This was the adversity that my family went through at that

time. They took those experiences of the 1930s and '40s on the chin but then allowed hardship to motivate them for greater things. It made them appreciate having their own and gave them a deeper appreciation for the land because they had all contributed their labor and struggle to keep it, so they had a vested interest. They had skin in the game, in their land and in their legacy. I once had a wonderful conversation with Gran's youngest son, my uncle Joe Carter Sr., who had an impeccable memory. I can recall his face lighting up with pride as he reminisced about the land and the sacrifices of his siblings. What he continued to share was that Gran had wanted to make sure she had a place for her boys to come back home to after the war. It was one of the important motivations that led up to Gran paying off a debt of $1,250 in 1945, thus securing the land for herself, her children, and all her descendants, up through my own sons today. In that process, she also secured the future of the family and ensured a legacy that would live on, well after her death in 1995. Sawubona Gran, Aunt Marjorie, Aunt Anna, Aunt Faye, Aunt Ruth, Aunt Mamie, Aunt Virginia, Aunt Bettye, Grandma Lucille, Grandma Rose, Aunt Arlene, Aunt Elizabeth, and Aunt Bernice.

Gran instilled in all of her children the responsibility of land and homeownership. Most of her daughters moved to the Washington, DC, area (and the rest to New York and California) and purchased homes. Before it was legal for a woman to apply for a credit card without a man's consent, before they had a right to vote, most of them owned a home, with or without a spouse. This ability, another anaerobic action, wasn't directly spoken about, but the fumes of the example they set got inside most of us and caused us to follow their leads.

Meanwhile, Gran made it a point to keep the farm in the family. Her vision was to make sure everybody had something. In 1978, she divided up the land among her children, giving each of her children six acres.

The amount of land lost by Black farmers and families in the last sixty years is, I dare say, a sin to our ancestors who made it their sole mission to acquire land between 1865 and 1965. I'm a benefactor of my ancestors' efforts and I refuse to let go of what was sacrificed by my grandfather and his brothers with their World War II checks.

THE HOMEPLACE

When I returned from Ghana in the summer of 2017, we had a family cookout at the farm. I hadn't seen the family in five years, although at every funeral or family gathering while I was away, my mother would call me and the phone would be passed around among thirty-five people so I could talk with them all. On this day, July 15, 2017, the land spoke to me, communicating that I needed to tell my family story.

As the sun was setting that day, a family meeting was held outside in front of my grandmother's house with all the immediate heirs of the property. This conversation revolved around what we wanted the future of the land and our legacy to be. How would we move forward? From this initial conversation, the idea of creating a family trust was planted, to be watered over the next few years.

During the COVID-19 pandemic, we held family Zoom meetings to communicate about the land, headed by cousin Mr. Greg Carter and the heads of each family, and we had a presentation from Ms. Ebonie Alexander at the Black Family Land Trust, who encouraged us to put the land in a trust to preserve it. (See "The Sixth Love Language—A Will" on page 80.) We understood and bought into her Black family land ethic, the idea that the greatest act of love you can give is to have a plan or a will when you die so your family doesn't have to fight or figure it out. She helped pushed the Uniform Partition of Heirs Property Act through the Virginia legislature in 2020. This state law assists in the prevention of involuntary loss of family land and the associated real estate wealth of that family land through a hierarchy of rights and remedies for tenant-in-common owners in a partition action.[1] Ms. Alexander's belief is that the more people you involve in ownership of a piece of land, the bigger probability you will lose it. If you have thirty-two heirs, with a great majority not connected to the land, they're going to sell it. The best way to honor your family and the sacrifices they've made is by putting land in a trust, and having the trustees make decisions. This is a way to establish, keep, and grow generational wealth.

It made sense to us—we've had this land for 100 years. Why sell it now? In 2019, there were approximately twenty-five heirs to our family property. (An heir is a person who is legally entitled to ownership of a property after the death of the person who owned it.) The heirs were scattered, some estranged from the family. And during the process, Uncle Joe transitioned, which meant his share of the land went to his children. This scenario could have become very complicated, very fast. But after numerous meetings, discussions, and deliberations we took a vote and everyone agreed to put the

THE SIXTH LOVE LANGUAGE—A WILL

The Black Family Land Trust (BFLT) works to keep Black families on their land in Virginia and several other states. Ms. Lillian "Ebonie" Alexander, BFLT's executive director, focuses on helping people understand the value of their land—to learn to see it as a precious asset, a performing asset. I met Ms. Alexander while doing similar work in agriculture in Virginia, and we have been good friends for the last several years. During a conversation, she told me, "In 2017, the World Bank said less than 30 percent of the world's population had clear title to their land. Do you know how blessed you are to have title to your land?"

The theme of value comes up over and over when I talk with Black farmers, chefs, and culture-keepers. It's an unfortunate but recurring fact that Black people don't always see the value in our own things, whether it be land, artifacts, food, culture, or our own bodies. This is a result of cultural conditioning and the anti-Black reality around the world.

Ms. Alexander is well aware of what's at stake. I asked her what she thinks about the future of Black land ownership and she said, "I'm scared. I see a scary and bleak future for Black farmers. We can't afford to lose another blade of grass or grain of sand." Through land loss, from 1925 to 2022, she told me, it's estimated that African Americans have already lost more than $326 billion in value.

The BFLT tries to prevent this through guiding orderly farm transitions, making use of easements, and helping farmers transition to new techniques. "It's not your granddaddy's farm anymore," she said. "We're looking at the use of seasonal high tunnels, no-till . . . things our granddaddies didn't know. One gentleman in North Carolina was a twenty- or thirty-year tobacco farmer, and he participated in the tobacco buyout. He reinvented himself as a freshwater prawn producer. He said once he got his easement and paid off his debts, he can lay his dead bones down to rest, knowing his land will always go to feed people."

Working with clients is an intimate process. "We are very hands-on with our work. We almost become part of the family because we're there 'til we get you to the finish line. All of our work is intergenerational."

The most important part of building that perception of value, Ebonie says, is to "*Make your land human*. I've made my land human; we can go back for generations and tell you things that happened on this land." Attaching those family stories to the land helps people build a sense of how irreplaceable it is. "You get a map of the land and put things on there. If you make your land human, it's like your children—it will get on your nerves but you'll never get rid of it."

She believes that people of African descent can and do have an innately harmonious relationship with the land. "How many times have you heard of the Egyptians trying to change the flow of the Nile? When the Nile overflows, what do they do? Get out the way. Who we are stands in reverence and harmony with who the land is. If we take care of that land, that land gonna take care of us, and my children, and my grandchildren." Sawubona Ms. Ebonie Alexander.

land in a trust. It was a six-and-a-half-year process from the planting of the seed to the seed bearing fruit.

The home place—Gran's place—is a security net, a safety blanket, a parachute for our family. It's something every family needs. That's what had brought me back from Ghana. It's our literal White House, and I'm grateful to the family for banding together to retain this part of our legacy.

SOCIETAL COMPOSTING

A compost pile might appear be a steamy heap of garbage, but if turned and cultivated properly, it can become a wholly different and useful substance. In the finished compost, all the richness of the leaves and twigs and cow, chicken, goat, horse, or rabbit poop is present but in a much different form. They all add to the mix, ultimately, to create fertilizer. There's a high probability that the transformation experience will be uncomfortable, though. When you're in the middle of the compost, you think "This pile is hot, moist, and it stinks, and I don't want to be here." But as Baptist church-going elders would bellow from the old wooden pews, "This too shall pass." Often, it was the energy and understanding of a grandmother or grandfather, a parent, an aunt, an uncle, or a cousin that helped make difficult situations mature into a rich compost to build on and grow upon. I picture them saying, "Let's give it a minute . . . turn it, turn it, turn it." With time and aeration, and intention, we ultimately end up with what's needed for growth. And the BS we encounter in our lives can speed up the composting process, making it churn faster.

That said, as a country, we need to do some serious societal composting.

When I teach and speak to groups of people, one of my goals is to toss the stereotypical tropes around African Americans and farming into the compost pile and explain why certain perspectives, ideals, and stereotypes are or are not valid. I have to literally break it down, or compost, these ideas for my audiences, giving them fertilizer to add to their existing knowledge and help them grow. For example, if I were to ask you, "What words come to mind when you think about farmers?" is "poor" the first word that you think of?

Why is the general stereotypical perception that people who work on a farm are poor? The owners of plantations never presented themselves as poor, but the farm *workers*, those who did the physical labor and many times the planning and preparation, were mainly African Americans and Indigenous Americans, and they started off as enslaved and indentured. This is one of the reasons you have the perception of farmers being poor. Societal BS put farmers in this situation.

Is "uneducated" another word one associates with farmers? Being educated was another uppityism that could get a Black farmer or family run out of their home and community or hung from a tree or lamppost. Keeping Black people financially poor and uneducated were the end goals of the Black Codes that African Americans experienced, which were very different from the codes we had internally in our communities to govern ourselves for our respect, dignity and safety. The Black Codes or Black Laws were laws made by the state or local legislators, specifically targeted toward the melanated population, that were interpreted and enforced by the local jurisdiction. These policies of law, order, and injustice first began to arise in the 1630s, from the Virginia House of Burgesses, the first Western-based legislature set up after the European arrival in the Iroquois

Confederacy. Virginia Act X in 1639 stated that all persons except Negroes were to be provided with arms and ammunition or be fined at the pleasure of the governor or council. This was the second time the Negro was mentioned in legislation post-1639 in the House of Burgesses. The first time was in 1630 for Hugh Davis, whose historic claim to fame was being the first white man accused and found guilty of having an intimate relationship with a Black woman; he was to be whipped for "abusing himself to the dishonor of God and shame of Christians by defiling his body in lying with a Negro."[2] These codes persisted across all the colonies and into the legislative bodies of the states as they were formed. Most of the laws have taken hundreds of years to break down and decompose in American society, and vestiges of them still leer in stereotypes and perceptions about non-white people in this country.

For African Americans, generations of physical, social, and cultural discrimination by the dominant elements of society have kept us from thriving. It's like the way industrial farmers grow a monocultural crop of corn. They plant nothing but corn, and nowadays it's likely to be a GMO variety so that they can spray 2,4-D, glyphosate, and a host of other chemicals to control exactly what grows. They make the environment as inviting as possible for one particular group or seed to grow and to thrive, and they use chemistry to prevent all other kinds of plants from growing. (Except for the "superweeds," which have outsmarted the herbicides and must be uprooted or physically attacked to be killed.)

The way this discrimination has played out is not just in personal relationships or local economies. It's happened through decades of federal policy that has buttressed and reenforced social conditioning. But looking deeper, the cultural rot of racism extends back hundreds of years. Enlightenment-era writers Francois Bernier, Richard Bradley, and Henry Home would flesh out and formalize concepts of race, noting physiological differences in human beings. Carl Linnaeus, the Swedish botanist and creator of the modern taxonomy system, influenced the European view of both plants and humans. Linnaeus claimed to have the ability to identify and judge plants by their native region, and implied that African plants were inferior. "A practical botanist will distinguish at the first glance," he wrote, "the plants of different quarters of the globe, and yet will be at a loss to tell by what mark he detects them. There is, I know not what look—sinister, dry, obscure, in

African plants; superb and elevated in the Asiatic; smooth and cheerful in the American; stunted and indurated in the Alpine."[3]

Linnaeus was also the first to classify humans in the kingdom of animals. He places humans in four divisions, initially based on geography and skin tone: *Europaeus albus*, *Americanus rubescens*, *Asiaticus fuscus*, and *Africanus niger*. In his tenth edition of *Systema Naturae* (*Systems of Nature*), released in 1758, Linnaeus added character and moral values to this distinction that, in turn, influenced societies for centuries. His description of the African species is more negative and derogatory than that of the other species. It notes their capricious governance and their neglectful, sly, and sluggish behavior, and goes into particular detail on African women's labia and breasts—based on sensationalized, exploitative, and assumptive accounts from travelers, explorers, and colonists and not on scientific observation. In this way, Linnaeus built the foundation for centuries of discrimination and stereotypes.[4]

Steeped in the highly contagious anti-African viral load of the seventeenth and eighteenth century, European men and women sought to standardize their perceptions in the arts, religion, and sciences. The ideas of Linnaeus and his kind became pollutants in the soil medium where Western ideology would plant its roots and grow. This was the poison known as scientific racism, and it metastasized into what I call ethnobotanical racism. It would taint every organism and philosophy that dwelled in its veiled layers.

Yet the baked-in racism of European and American culture takes many other forms, too, including some that seem innocuous—stuff made for kids. I remember going to the theater to see the movie *Dumbo*, in 1983 or '84, and also watching it in school as a treat. Have you tried watching *Dumbo* recently? In one musical number, Black workers erecting a circus tent sing about their illiteracy and laziness and wasting their pay. The movie also features a group of shuckin' and jivin' birds led by a character named Jim Crow, voiced by a white actor speaking in a perceived Black vernacular. When I share this with students, farmers, or participants in my workshops or lectures, everybody remembers it as just part of the movie, but not as being racist. I didn't recall it myself from seeing *Dumbo* as a child. I didn't notice the crows, shuckin' and jivin', or the crow named Jim Crow.

Up until the murder of George Floyd in the late spring of 2020, grocery shelves were still littered with branding subconsciously inspired by Linnaeus's stereotypical perceptions. Images of the Aunt Jemima mammy and the Uncle Ben butler or servant were finally turned to aerate these and other stereotypes of African Americans and our relationship to food. Stereotypes and racist, demeaning images had become so commercially acceptable in American society that it took a national call to action during a pandemic to execute a bit of performative change. Names of sports teams that demonized Indigenous nations or images that mocked African Americans became passé, not necessarily morally objectionable, just not the flavor of the month or what was trending in popular culture during the summer of 2020.

Stereotypes around the work ethic and intelligence of African Americans have also been perpetuated by laws and policy. Agricultural policies of the United States have often disadvantaged Black farmers and families. Whereas white farmers received government aid and were able to lift up themselves and their descendants through their work, especially in the 1930s and '40s, Black farmers weren't eligible for those programs and found themselves unjustly facing financial discrimination in a system designed to stagnate and potentially destroy their family agricultural legacy.

Subsidies for the Black community came later in a different form—as welfare and food stamp assistance accessed by women. The welfare system also led to a program that benefited dairy farmers (who were overwhelmingly white farmers) at the expense of the health of welfare recipients. A perceived dairy shortage in the 1970s led to the government stockpiling dairy to manage the dairy markets with price stability, stopping the downward fall of prices. The US government resorted to making cheese out of the surplus dairy, as a value-added shelf-stable product. President Jimmy Carter set up policies to subsidize dairy farmers, infusing $2 billion dollars into the industry. This encouraged farmers to, you guessed it, produce more milk, yet there wasn't a viable market for the milk or the cheese. By the early 1980s, the government had over half a billion pounds of cheese in cool storage. Secretary of Agriculture John Block informed newly elected President Ronald Reagan about this cheese problem. Following Block's advice, Reagan elected to give some of it away. The Temporary Emergency Food Assistance Program (TEFAP) distributed blocks to food banks and community centers, but the sometimes moldy, constipation-inducing government cheese wasn't a culturally appropriate food for all the demographics it was distributed to.[5]

The nuances of discrimination have not been easy to break down and dissolve in American society. Misogyny has been common in the United States, and the women's movements led by Susan B. Anthony, Ms. Sojourner Truth, and others are a testament to women's resilience. Ms. Truth also tackled the nuanced reality of Black misogyny. Another nuance that's rarely mentioned, except by a few scholars such as Dr. Tommy Curry and Dr. T. Hasan Johnson, is Black misandry. Misandry is prejudice, contempt, or dislike of men. Society has perpetuated this dislike and disdain for certain Black men. The Black Codes and Black Laws served as misandry protocol and normalization in the late 1800s. If Black men were moving around in public without a destination, they were defined and criminalized as vagrants. The White-Slave Traffic Act of 1910, now known as the Mann Act, made it a felony to transport women or girls across state lines for "prostitution or debauchery, or for any other immoral purpose." It was used to prosecute the Black boxer Jack Johnson when he drove his white girlfriend (later wife) across state lines.

As time progressed, and cinema became more mainstream in every American's life, Black cinema emerged. A compelling story of the reality of social welfare appeared in the movie *Claudine* in the early 1970s. The welfare system was a well-engineered design that made the woman the head of the household and incentivized her to be single. Black men were punished for doing their best, being natural providers for their families. At times, their provision was demonized, mocked, and sanctioned by the government, making it illegal for Black families to receive government aid, no matter income and circumstances, if a man lived in the home.

Racist perceptions and stereotypes should be thrown into the compost pile whenever they appear, not just subtly covered up. By discussing these issues openly, letting in the oxygen as we turn them over, we create something better—something that can actually be nourishing. I learned this growing up in Confederate-loving central Virginia, where racism was so baked in you kind of expected it. My high school history and sociology teachers, Coach Frye and Coach Raynes, were both Civil War reenactors, and they would occasionally wear their Confederate regalia to school. I respectfully challenged them, and we would have engaging, respectful, and thoughtful discussions around their regalia and positions of history. Their approach to my disagreement was never malice; they thought of it as history, not hatred. Because of that, I was able to engage in these discussions not from an argumentative or emotional perspective, but simply to explain the the history behind the automatic visceral reaction Black people might have to the Confederate flag.

I don't doubt my Confederate-sympathizing educators had some radically different opinions about certain accounts of history compared to me, but they were always up for respectful discourse. We still have conversations to this day. The art of composting can be enhanced by the practice of respectfully agreeing to disagree; the end result may not be agreement, but civility is the finished product.

This experience gave me a template, at a young age, for working with people who have opposing views, and a method for how to engage people, concepts, and information I disagree with. This is the work we all need to do if these racial ideas, many misinformed, we've been living with for so long are going to help us grow.

Composting happens naturally in lands throughout the world. In the Creator's world there is no waste; everything that's not been made by humans goes back into the earth, as nature intended. Antiquated thoughts, ideals, and behaviors can be transformed, as well. Our Black farmers carried many of these stories and experiences, and they chose to compost them in one way or another, aerobically or anaerobically digesting them, to sustain themselves. They modeled for me a form of alchemy to activate and preserve my dignity and respect for myself daily.

CHAPTER 5

FIRST LEAVES

THE CROPS THAT CREATED AMERICA

The fall of a dry leaf is a warning to the green ones.
African Proverb

When we talk about the formation of America, farming, and the history of people of African descent in this country, Virginia is at the core of it all. My family history goes back to the oldest plantation in Virginia, the Shirley Plantation, settled in 1613 in what is now Charles City.

The history of the Shirley Plantation—which is the oldest family-owned business in America—and its owners demonstrates the sheer wealth and power that some colonist planters accrued. At one time, for example, Robert "King" Carter, whose son John Carter married into the Shirley Plantation, owned almost 300,000 acres of land. Carter used his position in the House of Burgesses to snatch up the best land for his children and grandchildren. King Carter's descendants include other burgesses, a Virginia governor, an ambassador to Italy, and the Confederate general Robert E. Lee. Thirty-ninth president Jimmy Carter is a descendant as well.

Shirley Plantation at one time was the biggest farming operation in the state (see page 2 in the color insert). The Carters have owned it since the 1720s, with Hill and Shirley being the other names attached to the property over the years. Carter and Shirley are two of my family's surnames, as well. I believe some of my ancestors were enslaved at the Shirley Plantation, although as I've mentioned before, it is near-impossible for families like mine to find accurate records tracing our roots through this period. But my uncle Charles Shirley shared with me that the family used to receive correspondence from the Shirley Plantation back in the 1970s and '80s, inviting them down for cookouts. Gran and others would throw away the invitations; they didn't want to deal with it. Imagining the type of overtones or microaggressions that kind of gathering would have entailed, it was best to protect the emotions of the descendants of the enslaved as well as the emotions of the descendants of the plantation owners.

I try to go to Shirley Plantation every year; I've talked to some members of the Carter family, and it's been cordial. One thing the family clearly stated was, "We don't have any money." But I've never sought financial reparations. I just want to know the history.

Individual wealth acquisition, greed, exploitation, family, and nation building were the foundation of the plantation society—all constructed on the labor and knowledge of the Indigenous and the enslaved. This is how North American, Caribbean, South American, and European nations were built. At the same time the continent of Africa was robbed of its greatest wealth: its people and their creative genius. The final fruits of their lives in Africa and the first leaves of our existence in these lands of our initial captivity are generally not well known.

These first leaves, called the seed leaves or cotyledons, are just as vital to a newly migrated, kidnapped, or captured people as they are to a plant newly emerging from the soil. Cotyledons support the next stage of growth, the formation of the plant's true leaves, which are capable of photosynthesis and supporting a healthy, mature plant. The first leaves of the kidnapped African and captured Indigenous citizens was the agricultural knowledge of how to grow food and sustain themselves.

Were my ancestors enslaved at Shirley, and what did they help grow there? How did their skills help create that family's wealth—and the country itself, as those who profited from my ancestors' labor became leaders in the government and the military?

It all circles back to food. In the Americas, like most of the world, food production in the 1600s was hyperlocal, and most cultures revolved and evolved around their food, crops, and agriculture. The growing season in the colonies that started the North American colonial experiment spanned six to nine months. Indigenous peoples grew the three sisters—corn or maize, a bean/legume that fed the soil, and a squash.

Across the pond in the land mass that would be known as Africa, the food culture was as wealthy as the empires of Wagadu (Ghana), Mali, and Songhai.

The Wagadu Empire started in the eighth century and lasted through the thirteenth century, when a drought forced its complete decline. It occupied the area in what is presently southeast Mauritania and western Mali, from the Niger River to the Senegal River, and it was the first Western African empire to take advantage of the plethora of gold in the region. Sorghum, millet, yams, cassava, peanuts, legumes, and rice, along with a wide variety of fish and a cacophony of herbs and spices, made this region wealthy in food culture. This wealth would spread throughout the Western world for the next 500 years. A macro perspective of the crops that made America economically successful and viable to the British empire reveals that the plants that built the early colonial economy were sourced from Africa. A closer examination reveals that the people with the ancestral knowledge necessary to grow those crops were of African descent. The innovations that made some of these crops profitable and able to be grown on a mass scale are the work of Black inventors and genius teachers. And not only did African plants and people build this country, more than once they saved it.

RICE

Let's start with rice. It might not be the first crop you think of when you think of plantations in the South, but don't worry—cotton is coming along soon. Rice has a long history along Africa's western coast from Senegal to Liberia and Cote d'Ivoire, where people have been growing and eating native and Asian strains of rice for 4,000 years.[1] Indonesians who populated Madagascar around 1,200 years ago also brought rice with them and made it a central part of the Madagascan culture and economy. This was a major crop and trade item, and a staple ingredient in African dishes such as jollof rice, which is rice flavored with tomatoes, a multitude of herbs and spices, and a few secret ingredients. The name *jollof* is a variation of the Wolof, Djolof, or Jolof Empire and its people, who first created this delightful delicacy. The Mende people, who live in Sierra Leone and Liberia, grow rice as their most important crop, as well as a spiritual crop, and women are mostly in charge of planting and threshing the rice.[2] Rice mixed with palm oil is used as offerings at Mende ancestors' graves.

There's a legend about a ship from Madagascar blown off course in 1685 that landed in Charleston, South Carolina. In gratitude for aid given by the colonists, the ship's captain gave some seed grain to the colonial governor. That might not be the literal truth about how South Carolina's rice industry

OBRONI

The European enslavement of Africans was sometimes supported by a kidnapping process between warring factions of African nations. This brings questions to mind: Why didn't our African brothers and sisters come to help us, or fight to keep us in Africa? Why couldn't we unify and fight off this Eurasian scourge that devoured the foundation of African nations' growth and development?

To better understand some of the realities outside of the American or Western context, it's imperative to think using the African mindset of that time. Race is a relatively new construct for describing people and not every culture embraced skin complexion as a major identifier. The first definition of race in most dictionaries is an act of running a set course or duration of time, a contest of speed, a contest or rivalry involving progress toward a goal or objective. Thus, the most basic definition of race is a competition, but it's a competition where not every culture is actively or knowingly competing in the competition. From the inception of the concept of race, Europeans were competing to acquire power and control over resources, with skin color as the "uniform" that determines what team you are on.

The construct or conception of race isn't seen in the same manner or in other parts of Africa that I have visited, especially among their own citizenries. In Ghana, I realized that Ghanaians didn't see my skin color first, but rather used my character, mannerisms, and behavior as determining factors for how they perceived me. The most humbling feeling in the world for a Pan-African, Black nationalism–proclaiming, ultra-pro-Black man is to be referred to by a random stranger at an African market as *obroni*, a word in the Twi language that means "white man." He said it without any malice or disrespect intended; it was how people in Ghana saw me: as a foreigner. Although I was cloaked in what I thought was very stylish and comfortable West African fabric, I appeared foreign. The words I used to ask for things, my brisk walk, my smile, the way I wiped the sweat off my brow, which hand I used to give the merchant money, and my fast-paced central Virginia accent—all these signaled that while I may have been from there at some point, I wasn't of that country or continent now. My ways were that of a stranger, despite my skin color.

To be from the horizon, or from the far off place, is the literal meaning of obroni, which generally gets translated, as I mentioned, to "white man," or "the people from the horizon." This perspective helped me better understand that Africans rarely see themselves as having a Black identity, initially, but rather they are Akan, Ewe, Fulani, Ga, Mende, Tutsi, Amarah, Xhosa, Ndebele, Yoruba, Igbo, Mbundu, Bakongo, Fon, or whichever national ethnic and family origin group their family heritage is tied to.

An understanding of how cultural perspectives differed can give greater context to why many nations didn't unify around race during the colonial era. While there was a serious amount of resistance against exploitative kidnappers in the colonial period, those who resisted were not unified in their efforts by their skin color or being Black.

got started, but it's for certain that rice turned into a powerhouse cash crop. It was colossal in the early economies of South Carolina and Georgia. South Carolina's rice industry grew fiftyfold between 1700 and 1720.[3] By 1839, it had tripled again. And rice is still big today—it's the fourth-largest cereal crop in the United States.

Both African and Asian varieties of rice were grown in the colonies, helping to boost the economies of Charleston, Georgetown, and Savannah on the Southern coast. But it was African workers who, even more than the seed itself, were essential to rice production. Planters knew they needed agricultural masters who had experience not just in growing rice but also in building and sustaining nations.[4] All West African nations and empires had agricultural genius at their core, though most modern historians rarely acknowledge this. Advertisements touted slave ships carrying "a choice cargo of Windward and Gold Coast Negroes, who have been accustomed to the planting of rice."[5]

Many African citizens—especially women, the primary rice experts in their homelands—were turned into commodities and shipped across the Atlantic to power the new rice-powered economy of the American Southern coast and marshlands. As Leah Penniman shared in her book *Farming While Black*, some of them may have braided sacred rice grains into their hair before the journey, bringing with them the seeds of a beloved staple and spiritual crop along with their ancestral knowledge of how to grow and use it.

When you read about what it takes to grow rice, you can understand why colonial planters needed professional help. Rice grows in wet, marshy river deltas. Dikes are needed to hold water in the planting area, and a sophisticated irrigation system is essential for handling the fresh water that's mixing with the salt water during the tidal flows, and it has to be adjusted depending on the growth stage of the crop. In addition to this expertise, the master farmers of the Wolof and other dynamic West African empires knew how to polish the rice after harvest and even how to make rice-winnowing baskets from sweetgrass.

Not only is rice farming very taxing physically, it also requires intuitiveness and intelligence to grow it well, which is rarely considered, especially as related to enslaved or sharecropping agriculturalists. Everything was accomplished with hand tools alone—hoeing, plowing, milling. Workers stood ankle deep in mud, muck, and water while working the rice fields with the sun leaning on them from sunrise to dusk. Elders would sing soul-stirring, bone-chilling words such as "I'll be so glad when the sun goes down. . . I ain't all that sleepy but I wanna lie down. . . ."

Let's not even begin on what Ghanaians refer to as *ntumtum*, the dreaded mosquito that causes pain and irritation if you are lucky, and malaria or yellow fever if you aren't. In the Carolina and low-country bayous and marshes, it was also quite common to find alligators and venomous snakes that were not too discriminating about what they bit, chased, or tried to devour. Working in marshy water conditions, without what we would describe today as proper footwear, laborers suffered from foot diseases caused by fungi and waterlogged boots. Up to a third of these workers died within a year of arriving at the low-country rice plantations. Gullah chef and culture-bearer Mr. B. J. Dennis shared with me that at Gowrie Plantation on Argyle Island near Charleston, in Jasper County, 90 percent of the enslaved children died before their sixteenth birthdays. Historian Edda Fields-Black's account of Harriet Tubman finding enslaved workers toiling in the rice fields of South Carolina

Postbellum rice field in the South. *Photo from the Robert N. Dennis Collection of Stereoscopic Views, New York Public Library / Wikimedia Commons.*

at 3:30 in the morning during the Civil War is a testament to the maniacal unrelenting business of rice.

The death rates didn't really matter to the plantation owners—by the 1770s, rice was profitable enough that the rice a single worker could produce in one year was six times as valuable as that person's worth on the slave market. Even if a huge proportion of workers died every year at a plantation, that didn't mean the operation was a failure. This is the logic enslavers applied as they turned both plants and people into raw commodities.

The farm owners used this same sort of logic to deplete the soils in the low country along the East Coast, and rice production eventually shifted over to the Mississippi Delta, Texas, California, and to the rice capital of America, Arkansas. But rice had already left its mark on Southern culture, in the form of dishes including red rice, red beans and rice, rice and peas, pilau, hoppin' John, rice pudding, and jambalaya, developed by chefs of African descent, enslaved, indentured, and free, holding on to the memory and more importantly the taste of African cuisine. These are the American derivatives of jollof rice and other African foods.

Rice is about a $7 billion industry in America in the 2020s.[6] As a profitable commodity, it's been helping to build America for more than 300 years, and in its formative stage, its production was powered by the skills and back-breaking labor of Africans.

COTTON

Now we get to cotton—the king. King Cotton in the 1800s was the equivalent of the tech or oil industry today.

Read about the history of cotton and you'll find that most sources say cotton is native to subtropical regions on several different continents, and it was independently domesticated in the Americas, India, and eastern Africa.[7] The Kingdom of Kush, in eastern Sudan on the Nile River, may have domesticated the cotton plant 7,000 years ago.[8] Fragments of cotton cloth have been found in Meroe and Lower Nubia (ancient Sudan), and an Aksumite (Kingdom of Axum in Ethiopian region of Africa) king boasted about destroying cotton plantations there when he invaded the region.

In America, cotton seed may have been planted as early as 1556, in the Florida colony, and in Virginia in 1607.[9] Along the James River, where the first English settlements took hold, it was growing by 1616. Cotton *preceded* the enslavement of Africans in the colonies. It's symbolic of slavery because of its massive connection to the plantation system before the Civil War, but we don't think of cotton as being symbolic of Africa.

European greed created a necessity for an enslaved class of people just to cultivate this crop. Colonizers had an understanding that they could make a business selling cotton, but they didn't have the labor force to grow it on a scale that would be profitable. The Europeans who migrated to the Americas were primarily people looking for opportunities to live in their concept of freedom, and they weren't necessarily the best farmers. As historian Howard Zinn recounted in *A People's History of the United States*, the early colonists almost starved to death in 1609–10.[11] This time period was labeled the starving times, with the colonists "beefing" or "having opps," that is, not playing well together, with the original residents, citizens, and preservers of this land.

The colonists grew more tobacco than cotton—cotton production stayed relatively small until the late 1700s, but then the picture changed. The British wanted American cotton for their textile industry, which was growing fast due to the new technologies of the Industrial Revolution. The first US citizens saw an opportunity to steal the vast open lands of the Western frontier for cotton production with a built-in management system (the surviving colonizers) and, most important, the enslaved labor that the colonies possessed, the solid stem of the colonies and the nation.

PATENTED PLANTERS

Mr. Henry Blair, a free Black farmer in the early 1800s who lived in Maryland, is not as well-known as some of the famous Black innovators in agriculture, but his inventions have had an immensely important effect on modern planting methods.[10] Mr. Blair invented a mechanical corn planter in 1834 that made it faster and easier to sow corn seed and also improved germination rate, thus increasing yields. His invention was like a wheelbarrow that could be filled with seed corn, to be dispensed at an even rate into furrows. Rakes attached behind passed over the seeded furrows and covered them with soil. Some think that he was illiterate, but nonetheless he received a patent for this device, making him, at the time, only the second African American to hold a US patent. Two years later, he patented a cotton planter. Mr. Blair's invention is the prototype for every seed planter device worldwide.

About 700,000 enslaved people lived in the United States in the early 1790s, when Eli Whitney invented the cotton gin.[12] With that invention, the industry was suddenly positioned to spread like COVID-19, literally around the world, with similar deadly effects for those not fully vaccinated against whiteness. The enslaved population grew at a breakneck pace. Over 3 million enslaved people were in the United States by the time the Civil War started, and meanwhile the country had become the world's largest producer of cotton. In the states that produced the most cotton, half of the residents were enslaved. Their "masters" were some of the richest people in the country.

By the end of the Civil War, the cotton industry employed or enslaved roughly 20 million people around the world. Cotton fabric had become the standard for people in China, India, and Africa, as well as in Europe and North America. Cotton was critical—the fabric of our lives, like the commercials used to say.

The cotton plant had a greater impact on the lives of people of African descent than we will ever understand. I can recall hearing relatives and elders

Table 5.1. The Philosophical Aspects of Cultural Difference by Dr. Edwin Nichols

Ethnic Groups, Ethnic Worldview	Axiology
European Euro-American	Member-Object: The highest value lies in the object or the acquisition of the object
African African American Latino/a Arab	Member-Member: The highest value lies in the relationships between persons
Asian Asian American Polynesian	Member-Group: The highest value lies in the cohesiveness of the group
Native American	Member-Great Spirit: The highest value lies in oneness with the Great Spirit

Adapted from Bruce A. Jones and Edwin J. Nichols, *Cultural Competence in America's Schools Leadership*, Engagement and Understanding (Information Age Publishing, 2013), 37.

tell someone, generally not in kindness, "Get your cotton-pickin' hands off of that!" We have had a love-hate relationship with cotton. It wasn't the plant's fault that so many of us were commodified and abused in order to grow it. The plant was also exploited and commercialized. Its mass exploitation led to our mass exploitation. We were both victims of the same mindset, attitude, and spirit. Can you or should you blame something that's part of the commodification of you? Do we blame the belt for our childhood spankings, whippings, or beatings? (For those of us who are familiar with such corporal punishments.)

If not cotton, they would have found something else to commodify. Devoted and dedicated colonizers put on themselves the moniker of conquistador, entrepreneur, explorer, missionary, or tourist, and they traveled all around the Earth to find people, places, resources, and philosophies to colonize. Trying to create as much profit as possible with as little effort as possible was the primary goal for many of them, in addition to seeking a new life. When you have this ideal of cheap labor producing profitable and popular goods, that becomes the foundation of capitalism.

Epistemology		
Applied	**Pedagogy**	**Methodology**
One knows through counting and measuring	Parts to Whole	Linear and Sequential
One knows through symbolic imagery and rhythm (function)	Whole Holistic Thinking	Critical Path Analysis
One knows through transcendental striving	Whole and Parts are seen simultaneously	Cyclical and Repetitive
One knows through reflection and spiritual receptivity	Whole is seen in cyclical movement	Environmentally Experiential Reflection

Axiology is the study of values, and table 5.1 presents a summary of the brilliant, ahead-of-his-time work of clinical psychologist Dr. Edwin Nichols. In his summary, we can see the key differences in the way various cultures approached the precolonial world and determined what was most important to them. Dr. Nichols argues that African cultures were not based on exploitation because of the natural abundance of food and resources they possessed and easily shared. Exploitation by European nations arose from the fact that there was not as much agricultural abundance and food abundance in Europe, where the growing season is shorter. This lack of abundance affected the culture of that region, where the conception of time was much more critical, such that systems had to be created and instructions had to be followed precisely and on time. There was a literal deadline: if you planted something too late or not at all, you could die in the winter. These environmental factors also created a need for measuring and counting and prompted the concept of efficiency. In Europe, the culture was not based upon making sure everybody had enough. Instead, there were the haves and have-nots.

You were the guy who lived in the castle, or you were the serf who worked for the family within the castle. Sawubona Dr. Edwin Nichols.

Contrast this to the African values that emphasized trade and exchanging of cultural information, instead of getting wealthy and exploiting others in the process. In West Africa, you can pass a king's palace and not realize it. The king's house usually looks like everybody else's house in the community. This is the house where the king lives and therefore it's a palace. It's not the size of the house that makes it a palace (this is in contrast to the king's estate in Europe, where you might have to cross eight miles of lawn and a moat to get to the house).

The idea of the fiefdom, with the serfs subservient to the feudal lords is a mindset that accepts the validity of unequal control of resources, and it has become the basis of capitalism around the world. It's different from the mindset of many Indigenous cultures, who logically consider that if you didn't have to pay for something, you don't have the right to sell it. You live within your means, working as a steward and not necessarily as an owner. There's an understanding that if I'm provided for, we're all provided for. I am you and you are me, the African principle of Ubuntu.

But in America, the cotton crop—grown and picked by Black workers—fueled the growth of incredible wealth in the hands of a few. And the house of those highly fortunate few, incidentally, had those same large lawns in front of them. Because of King Cotton, the plantation class gained the ability to imitate the lifestyle of European nobility.

It's very possible that my ancestors on the Shirley Plantation were involved in growing cotton—and cotton was a cash crop. Right down the river from Shirley Plantation was Sherwood Forest Plantation, the home of the tenth president, John Tyler. Just recently, Tyler's last living grandson, Harrison Ruffin Tyler, died at age ninety-six in May of 2025. He'd been a preservationist and businessman with great privilege and resources.[13] Harrison Tyler and I both have family roots in the same area, but his ancestors owned the land and my ancestors were owned. It's a microcosm of the different paths that agriculture has led families and legacies upon.

Cotton played a major role in soil degradation, and inadvertently played a role in family growth or deterioration. Legacy formation for many had a direct correlation with their relationship to the cotton plant. Cotton was not broadly grown when Turtle Island became the United States, but production exploded after the War of 1812. The German- and British-based Baring family became a leading force in banking and the emerging cotton textile industry after the death of their patriarch Sir Francis Baring in 1810, who left 600,000 pounds of cotton, valued at about $50,000, as an inheritance to his offspring. (This translates to about $1 million in today's money.) In 1812, the family invested 6,000 British pounds in a cotton export house in New Orleans, Louisiana, which provided significant seed money for a global industry. By 1845, the influential family was importing some 104,270 bales of the new white gold—the production output of 70,000 thousand enslaved persons. The sale of cotton made many wealthy, and the Barings rode the wave of their wealth until 1995 when their main institution, Barings Bank, worth $1.3 billion at that time, defaulted. By 1860, cotton mills in northeastern United States employed some 400,000 people. The enslaved population of the United States was 3.2 million, and 1.8 million of them were involved in growing and processing cotton with no benefit to their legacies or security. Across the pond, one-sixth of the English population, approximately 4 million people, were involved directly or indirectly in the cotton industry. About 20 million people around the world at the start of the US Civil War were employed directly or indirectly in the cotton industry, making it the second largest global commodity after its enslaved African cultivators.

Historians, such as Dr. Ivan Van Sertima in his book *They Came Before Columbus*, provide greater detail into the journey and African origins of cotton and its impact and influence on the cotton grown in the Americas. Cotton has influenced economies, people, nations, soil, and marketing. This African crop has clothed and covered the world a few times over. Cotton was king as a commodity crop, and those who grew cotton and made it prominent were commodities. It became the first global commodity before oil, rubber, and social media, and it serves as a key ingredient along with linen in the literal production of physical United States currency bank notes. African crops are the basis of the US dollar, and for making the pockets of the jeans it is placed in. For the last two centuries they have been the fabric of all of our lives.

Black cotton farmers have been on a sharper decline than the broader Black farmer population. However, farmers like Mr. Julius Tillery (see "Julius Tillery" in chapter 9) are working to reverse this decline, and challenge the perception associated with African Americans and cotton.

INDIGO

There's a decent chance you are wearing jeans as you read this, or if you aren't wearing a pair, you own a pair or two. Americans buy 450 million pairs of jeans a year.[14] Jeans long ago became the uniform for farmers and many Americans in general. But most people don't stop to ask where the classic look for jeans—blue cotton stitched with yellow thread—came from. The answer's in another crop important to the Africulture framework: indigo.

Indigo is the plant dye that gives denim its iconic color. Indigo dye is extracted from plant species that belong to a big family of tropical and subtropical plants. Africa is one of the places where indigo is native, although it's named indigo because it was first noticed by European traders while it was being cultivated in India. West Africa also had a trade in fabric dyeing and textiles that was established well before Portuguese colonizers arrived in 1342 and noted the skilled local craftsmanship.[15] Dyes would have been used for coloring fabrics such as mudcloth, adding adinkra symbols on clothes, and for creating artwork. Mudcloth, a cotton-based fabric based out of the Mali region of Africa believed to have originated around 1100 CE, is traditionally dyed with fermented mud. Some believe that the Dogon nation started to dye mudcloth with indigo in the early 1400s and that the Peruvians introduced indigo into their fabrics as early as 4000 BCE. There is evidence of indigo use in Kemet, or ancient Egypt, as well. Like many of the great cultural and agricultural feats of Africa, the development of indigo dye there is rarely acknowledged.[16]

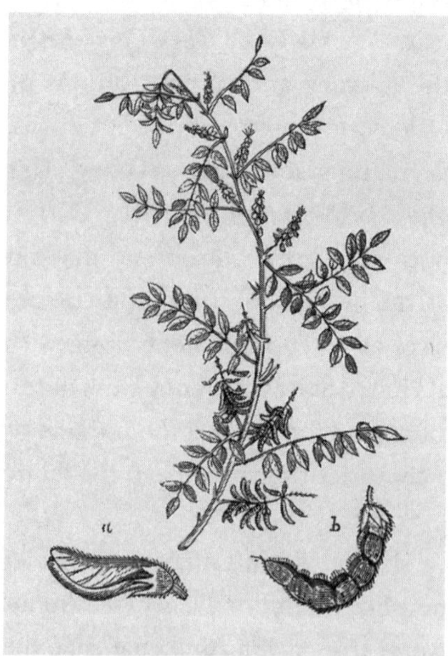

Indigo plant. *Photo by ilbusca / iStock.*

Indigo was a sign of wealth, fertility, and abundance in Africa. It was used for transforming cotton into deep blue cloth, notably by the Mandinka in Mali. As part of their trade, during this time period, Mali traders introduced the world to many of their cultural ways, customs, and traditions. All the accounts of Mansa Musa's Hajj describe the adornments—Mansa

Musa was not known for going to Mecca naked. The entourage was dressed very nicely in fabrics that would have been dyed with indigo.

In the colonial Americas, indigo was first grown in the Caribbean, and indigo seeds came to South Carolina as early as 1670. But indigo was difficult to grow and process and wasn't a crop that English settlers were familiar with. The industry didn't develop until 1740, when an eighteen-year-old plantation owner, Eliza Lucas Pinckney, started growing indigo with the agricultural innovation around her—dozens of enslaved scientists under her purview. Her father was the governor of Antigua, where the seeds came from.

Eliza gets credit for growing the first successful indigo in the colonies, and she did experiment with cultivating the plant from seed and extracting the dye. But who was doing the actual work? The enslaved people on her plantation. And because of that long tradition of indigo in West Africa, many of the people stolen from there and brought here arrived with generational knowledge of how to grow and process indigo. In *Black in Blues*, interdisciplinary scholar Dr. Imani Perry writes that an enslaved Black man from Guadeloupe or Saint-Domingue was the one who taught Eliza to cultivate indigo.[17] (I admire Dr. Perry's perspective, acknowledging that she treats folklore like documents when she is researching. I'm learning to do the same thing, as our folklore has historically been how we shared the contributions and significances of our cultures. Sawubona Dr. Perry.)

To grow indigo required the labor of about fifteen people per fifty acres. And once that crop was harvested, you needed another twenty-five skilled people to turn it into dye.[18] They had to be able to build and maintain all the vats and machinery for processing, manage the fermentation of the leaves, agitate the liquid in the vat in just the right way, and correctly dry the finished product to pack into barrels for export. At the peak of the industry in South Carolina in 1775, 1 million pounds of indigo was traded to Britain.

Indigo production and the slave trade grew hand in hand. The need for labor to produce indigo was great enough that South Carolina ramped up its importing of enslaved Africans. Moses Lindo, a British merchant turned indigo inspector stationed at the port of Charleston, owned a slave ship that transported human cargo from Barbados to Charleston. Lindo started actively in the indigo trade in 1752 and increased the volume of trade more than fourfold, so by 1772 he was shipping more than 1 million pounds of indigo per year. The increase in production of indigo was parallel to the

STONO'S FREEDOM REBELLION

The State of Georgia was established as a buffer zone between the majority African and Indigenous population in South Carolina and the Seminole nation to the south, where a number of African people had gone after escaping slavery. In 1739, in the low country region of the colony of South Carolina, a warrior named Jemmy, or Cato, led the Stono Rebellion. Mr. Jemmy was believed to be from the Kingdom of Kongo and organized as many as eighty fellow enslaved captives, mostly from the Kongo/Angolo region. They killed twenty of their captors and planned to kill their way through Georgia to Seminole territory where they would be free. (Seminole means runaway, wild ones, or separatists. They were composed of the Creek nation and other Indigenous or African individuals who refused to be colonized.) In response, South Carolina passed the Negro Act of 1740, which made it illegal for enslaved Africans to assemble in groups, raise food, earn money, move abroad, or learn to write. Landowners were permitted to punish "rebellious" enslaved Africans in any way they liked, including death. Sawubona Mr. Jemmy.

increase of kidnapped Africans forced into enslavement, and by 1860, enslaved Africans outnumbered free whites in South Carolina. [19] Next door, in Georgia, slavery was outlawed until 1751, when Georgians decided they couldn't afford to stay out of the indigo game any longer. They legalized slavery, and off they went.

Indigo was a "first leaf" crop in America that assisted in catapulting colonial America into a global agricultural giant, and back in Africa nobody could have foreseen how it would be used in the manner it was. The leaflet doesn't appreciate how it's the start of something else, as it's just functioning to provide resources for the plant. For a commodity, the focus is on exploitation and abundance of quantity—"We need to have these sold by XYZ time." In a trade mentality, it's, "My family made six of these, what can we get in exchange for those?"

I've heard that in West Africa, the blue tips of people's fingers could have been a badge of honor and sense of pride in a viable, honorable occupation. "I come from a long legacy of indigo growers; this is what we do." Other people would have seen the blue fingers and said, "I know what you do, you have a place in our society." Indigo fingers, sometimes called Elukami indigo fingers, have a spiritual meaning and connotation in spiritual practices like hoodoo. It's used in some washes to clean and purify floors and surfaces, to ward off unwanted energy, and to remove negative energy.[20] But after a while, the blue fingers became a sign that people disdained because of the hard work required to make indigo and the lack of pay for their efforts. Blue fingers were a clear, long-standing sign of the horrors of the enslavement period.

The indigo industry was important for about fifty years in South Carolina—for a while, it was second only to rice. It ended because the American Revolution made it impossible for American growers to sell their indigo dyestuff to their main customers, the British, and they couldn't find another market that big. Just in time, here came the cotton gin, and many plantations made the switch.

PEANUTS

After the end of slavery in the United States, during Reconstruction and the rise of the American apartheid system, Black Americans had to contend with the turmoil of society being reorganized even while racism and discrimination continued to impact their lives and opportunities. By the end of the 1800s, many of the emancipated and their descendants were working to make a living on some of the poorest land in the country.

Black people had been systematically funneled onto sites that had poor and depleted soil. Even if you owned land, you may have had to do some sharecropping, because the only land you were given was far from the best or too small to support the family. Prime real estate was always given to white landowners. Land with heavy clay soil, swampland, land where soil had been severely eroded by cotton or corn monoculture—that's what was begrudgingly given to Black landowners in most circumstances. And sharecroppers often ended up owing their "landlords," the owners of the plantations, money at the end of a growing season for unfair or contrived expenses, instead of turning a profit.

Crop rotation was not part of the growing system on cotton plantations. White settlers had run the Indigenous nations off their sacred grounds, not bothering to learn any of the Indigenous techniques for growing crops without ruining the land. The Indigenous peoples' three sisters approach of growing corn, beans, and squash together, for example, could have saved the soils. This approach created a symbiotic growing environment, where the beans provided nitrogen for the corn and the corn provided a trellis system for the squash and/or the beans, and all fed and added to the soil. But the farmers of European origin operated by planting monocrops as heavily as possible, then labeling the soil bad or worn out, and moving on to another plot of the land to repeat the process. That became a challenge in the South as the supply of arable land ran out. Growers found themselves competing not just with other white people but, after emancipation, with African Americans. This competition is part of what led to lynchings and raids on Black communities—in addition to being crude, evil, terror-inducing racist persecution, these actions were also a way to acquire land cheaply if not flat out free.

Dr. Carver and His Miraculous Goobers

This was the scene when Dr. George Washington Carver began his work at the Tuskegee Institute (now Tuskegee University) in the 1890s. Nowadays, Dr. Carver gets relegated to being the "peanut guy" every year during Black history month, but his accomplishments also included leading the way in conservation, organic agriculture, cover cropping, crop rotation, value added products, and even the creation of the first biofuels.

Dr. Carver had spent his life pursuing knowledge and excellence. He was born into slavery, in Civil War–era Missouri, and as an infant was kidnapped along with his mother during a slave raid. His mother was never heard from again and Carver was returned to the plantation. He ended up spending most of his childhood in the home of the Carvers, the family who enslaved him.[21] There's no way to prove it presently, but it's assumed in some circles that Dr. Carver had been castrated as an enslaved person. He worked as a house servant; it wasn't unheard of for boys in that situation to be castrated to keep them from desiring the white women of the house. Dr. Carver was noted for having a high-pitched voice and no known intimate desire for the opposite sex. This is an area of dispute, but I've heard from the likes of both the scholar, activist, and comedian Baba Dick Gregory and the Carver expert

Dr. Carver, center front, with colleagues at Tuskegee. *Photo by Frances Benjamin Johnston / Wikimedia Commons.*

Charles Micheaux, that Dr. Carver was castrated in his preteens or teens. If this was the case, this is an example of cutting off the first leaf in metaphorical form, and the plant being resilient enough to become the great species that it was bound to be, despite its potential mutilation.

Dr. Carver was recruited by another legendary figure, Dr. Booker T. Washington (see "Dr. Booker T. Washington" on page 113). Dr. Washington asked Dr. Carver to head up the then new agriculture department at Tuskegee. At the time, Dr. Carver was the only African American who held a graduate degree in agricultural science. For forty-seven years, Dr. Carver filled his role proudly and responsibly, placing the intellect and genius of his department at the center of the world stage of agriculture numerous times.

Dr. Carver's knowledge saved American agriculture time and time again. He wasn't just the peanut guy but an agricultural savant, thinking outside the box, and sometimes outside the soil of traditional agricultural thought. His reasoning for promoting peanuts as a crop was calculated: to improve both the soils of the South and the economic prospects of Black farmers. In the late nineteenth and early twentieth century, those farmers faced the challenge

Dr. Carver with a student. *Photo by Clifton Johnson / Wikimedia Commons.*

of coaxing plants to grow in the nutrient-depleted soils that resulted from excessive cotton growth.

Dr. Carver had observed that peanuts were a great cover crop because they're leguminous. Legumes are crops that can form a partnership with bacteria in the soil to transform nitrogen from the air into nitrogen compounds that plants can absorb and use for growth. This is called nitrogen fixation, and thus, growing peanuts could restore nitrogen to depleted soil.

Not only that, but they could also provide a good protein source in Southern diets. Yet peanuts—or goobers, as they were called at the time—weren't a very valuable crop, and their use in the United States was very limited.

The goober, or peanut, *Arachis hypogaea*, has a long, complicated, slightly confusing history. Per genetic research, the Andes Mountains of Bolivia are thought to be the home of the plant now grown as the modern peanut plant. The Aztec referred to it as the ground, or earth, cocoa bean or *tlālcacahuatl*; in China it is called *lo hua shen*, which directly translates to "born from flowers fallen to the ground." (See page 9 in the color insert.) I once thought the peanut had a doppelganger from the other side of the world—the groundnut (*Vigna subterranea*), which is a genetically different crop. (This confused me until I read *Slaves for Peanuts* by Ms. Jori Lewis.) Their leaves and plants look similar, they both can grow in shells, and I have a strong feeling they both spoke to Dr. Carver fluently.[22]

In Ghana, I ate thousands of roasted groundnuts, the perfect healthy snack, sold for 50 pesewa (about a dime) and packaged in little clear plastic baggies twisted at the top. For many years I thought groundnut was just another name for peanut, like using cowpea to refer to what I knew as the black-eyed pea. In many places the names have grown to be synonymous with the peanut; however, a few cultures who understood the difference had different names for each. The Wolof named the groundnut *gadianga* and the peanut *gerte*, but the Mandinka referred to both as *tiga*. The names goober and peanut derived from variations of African names of the Kongo and Kimbundu nations: *nguba* and *mpindu*. In Jamaica, it's noted that enslaved farmers called them *pindalls* or *gub gubs* and that they were excellent for the soils of Jamaica. In North America, the first names of this plant were pindars or goobers. Like most foods from Africa, peanuts were considered slave foods, foods for the poor, of low value and low worth, both in status and nutrition.

Surprisingly enough, it was war that brought these soil-improving nutritional powerhouses into the larger American dietary lexicon. During the Civil War, food rations were critical. Many battles were won or lost over the ability to provide soldiers the sustenance and the energy they needed to fight. General William Tecumseh Sherman, who utilized the United States Colored Troops to guard his supply lines in Georgia, was known to order fields burned and supplies destroyed to hamper and weaken the Confederate army. Peanuts became a wartime provision for many soldiers on both sides; boiling peanuts preserves them for several days without refrigeration.

After the war, the peanut became more respected as a food commodity but was not abundantly grown. It was nowhere near cash crop status. Some Black farmers in the apartheid South at the turn of the century couldn't afford to plant cover crops; they needed the income from a cash crop every year.

The solution was to create a market for peanuts, so that it could fill the role of both cover crop and cash crop. Dr. Carver is famous for creating more than 300 value-added peanut products—a multitude of peanut recipes and concoctions including coffee substitutes, milks, ice creams, imitation meats, and dozens of household and industrial products that included paper, paints, oils, and shampoo.

I wouldn't be surprised if Dr. Carver's knowledge of peanuts came down from his ancestral lineage, which may have been from a groundnut-growing area. It's widely believed that the peanut plant spoke to him. The following quote, often attributed to Dr. Carver with no official source, signifies the spiritual attunement and acuity that Dr. Carver possessed and exhibited in his daily life and research: "When I was young, I said to God, 'God, tell me the mystery of the universe.' But God answered, 'That knowledge is for me alone.' So I said, 'God, tell me the mystery of the peanut.' Then God said, 'Well George, that's more nearly your size.' And he told me."

Not surprisingly, this agronomy savant did the same for soybeans (another nitrogen-fixer), sweet potatoes, and pecans.[23] The father of value-added production, Dr. Carver showed farmers how to earn a living by making viable products out of what nature and the land produced. Virginia and Georgia became the peanut capitals of the world, and farmers there were able to restore the productive quality of the soils. This crop, coupled with the genius of Dr. Carver, reintroduced Indigenous crop rotation practices back into the Turtle Island agricultural landscape.

Dr. Carver was a man of faith who believed that spirituality went along with science in a unified quest for the truth. He would go outside in the early morning and just listen, and the Creators transmitted information through the observations of the plants he grew. As a young person he was an artist, and his artistic excellence rivaled his agricultural excellence. At the 1893 World's Fair in Chicago his painting *Yucca and Cactus* was awarded an honorable mention prize.[24] He turned that into an understanding of how to listen to the herbs, to nature, and become an agricultural genius just by allowing himself to be a vessel.

"I love to think of nature as an unlimited broadcasting station, through which God speaks to us every hour, if we will only tune in," Dr. Carver wrote.[25] Higher levels of spirituality in many belief systems focus on being quiet, observing, and listening rather than proselytizing your beliefs. I was taught early in my spiritual path to sit down, shut up, and listen. Listening to hear and learn and not to respond is an Africulture practice. As the old adage goes, you have two ears and one mouth, you should listen twice as much as you speak.

Dr. Carver, turning his attention to the soil. *Photo by Frances Benjamin Johnston / Wikimedia Commons.*

Dr. Carver not only listened to the plants and the natural world, he also listened to the farmers around Tuskegee and the entire state of Alabama. In my eyes, this listening nature made him the father of the

A Jesup wagon. *Photo by USDA / Wikimedia Commons.*

modern agricultural extension services. He designed a demonstration lab on wheels, called the Jesup wagon (named after the sponsor of the initiative, New York banker Morris Jesup), and rode out to do what we today call outreach. Dr. Carver and his staff went to rural and remote areas of Alabama, listened to the underserved farmers' needs, and came up with solutions.

Their work evolved into the first agriculture extension service in 1906 and inspired the country's first extension agent, Dr. Thomas Monroe Campbell, who operated a "moveable school." Dr. Carver also created the Carver bulletins—several dozen pamphlets on everything from tomatoes to hogs. He set up fairs and short courses for farmers, too, promoting self-sufficiency and self-improvement. In agricultural terms, this is what we now call technical assistance. Africulture and many other agricultural support institutions copy the model of Dr. Carver and Tuskegee to provide outreach to small, beginning, minoritized, socially disadvantaged, historically underserved, and specifically Black farmers.

The genius and innovation of Dr. Carver and the Tuskegee agricultural machine doesn't stop there; it's just getting started. Dr. Carver invented the first biofuel, some 100 years before President George W. Bush mandated the addition of corn-based ethanol to automobile fuel after 9/11 as a way to support corn farmers. As the auto started to burgeon in the 1920s, Dr. Carver envisioned creating fuel oil from peanuts. Growing fuel could give Southern Black farmers definitive income no matter what the economic climate.

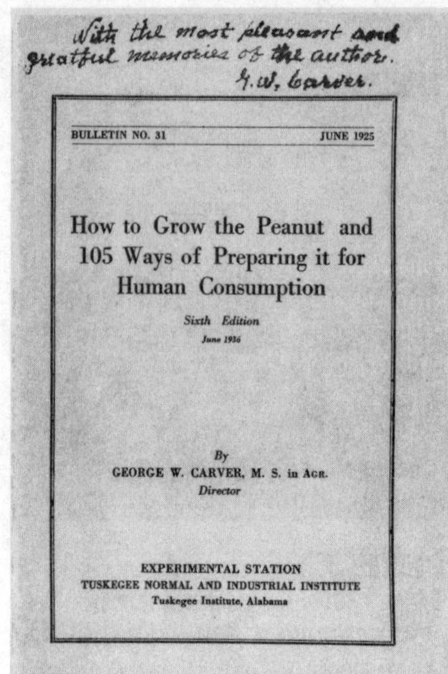
With the most pleasant and greatful memories of the author.
G. W. Carver.

BULLETIN NO. 31 JUNE 1925

How to Grow the Peanut and 105 Ways of Preparing it for Human Consumption

Sixth Edition
June 1936

By
GEORGE W. CARVER, M. S. in AGR.
Director

EXPERIMENTAL STATION
TUSKEGEE NORMAL AND INDUSTRIAL INSTITUTE
Tuskegee Institute, Alabama

A Carver bulletin on peanuts. *Photo by the National Agricultural Library.*

Rudolf Diesel, who invented the diesel engine (not the fuel), showed a version of his creation at the 1900 World's Fair that ran on peanut oil. He was in communication with Dr. Carver about that oil, which would have catapulted the African American economy if it had become the standard fuel for diesel engines. We would have seen this pave the way for farmers to stay in agriculture going into the Depression. Can you imagine where the African American community could have gone, and what it would be, if its economic growth had matched the pace of the America gas and auto industry? There might not have been a Depression if the efforts of Dr. Carver and Mr. Diesel hadn't been thwarted.

Diesel was an advocate for vegetable-based fuels.[26] But Mr. Diesel was found floating in the English Channel in 1913 after disappearing from a passenger ship. No one knows why; he took dinner on board the ship and was never seen alive again.

Diesel wasn't the only inventor who knew, respected, and sought to collaborate with Dr. Carver. Henry Ford tried to solicit him on numerous occasions. They collaborated on lab research in Michigan, inventing products from soybeans, and they discussed other plant-based materials as alternatives to metal and rubber. I don't think Dr. Carver got a lot of compensation for this, though Ford did offer him numerous jobs, which Dr. Carver refused. Ford did honor him by naming a lab after him and having a replica of Dr. Carver's log cabin birthplace built in Dearborn, Michigan.[27]

Dr. Carver died in 1943. He had truly lived out his purpose. "It has always been the one ideal of my life," he wrote, "to be of the greatest good to the greatest number of my people possible and to this end I have been preparing myself for these many years, feeling as I do that this line of education is the key." Booker T. Washington and George Washington Carver were the first two Black people whose images were minted on an American coin. They were shown together on the half-dollar in the early 1950s. Sawubona Dr. George Washington Carver.

Dr. Booker T. Washington

Without Dr. Booker T. Washington, there never could have been a Dr. Carver.[28] Dr. Washington was an immensely important first leaf on the vine that grew up to bear the fruit of many of the agricultural geniuses of the early twentieth century in the United States. Washington was born into slavery in Franklin County, Virginia, and he was eleven years old when the Civil War ended. His mother moved the family to the newly created state of West Virginia. At age sixteen, looking for education, he walked himself 500 miles, to southeastern Virginia, to enroll in the Hampton Institute. In 1881, he became the first principal of what we now call Tuskegee University in Alabama. He put excellence first. Second, he put understanding of where you are and what you need to do to survive and prosper, along with what he referenced as character-building. His dedication to education, community building, and Pan-Africanism was phenomenal.

Dr. Washington has been a major influence on my philosophy ever since the summer of 1994 or '95, when I picked up his book *Up from Slavery* and became intrigued by the story of his life. That book is an essential inspiration for this book you are reading now.

Tuskegee was built by the students and community of Tuskegee and Macon County, Alabama, making their own bricks, growing their own food, sewing their own sheets and clothes, and building their community. At the 1895 Cotton States Exposition, in his "Atlanta Compromise" speech, Washington encouraged African Americans of that time to become experts in the fields of agriculture, mechanics, domestic service—all the vocational trades. He used the metaphor of "casting down your buckets where you are," by which he meant, do what you need to do where you reside in the society: what you need is right in front of you. His detractors of the time said this was an accommodationist message that didn't encourage or support equal rights for all, but rather pigeonholed Blacks into the areas that they already occupied in terms of the vocational trades and arts. Dr. Washington, however, had a vision; he had achieved at Tuskegee self-sufficiency, making himself and his institution invaluable to others, and increasing the wealth of the community. Tuskegee became an internationally renowned name in the early twentieth century, and Dr. Washington began to see a greater vision and connection with Africa and its relation to melanated peoples in the United States and the Caribbean.

At times, Washington denigrated Africa, but at other times he uplifted it. He never visited Africa himself, and it's tough to get a full perception of what Africa means until you're fully there physically.

Booker T. Washington. *Photo by Francis Benjamin Johnston / Wikimedia Commons.*

Still, in 1901, Dr. Washington had several of his Tuskegee students attend the Togoland Cotton Expedition. Washington wanted to create a Tuskegee-style university and vocational training center in Togo, but it did not come to pass. He was also influential in getting the African Exclusion measure removed out of the 1915 US immigration bill. This proposed law, which would have prevented people of African descent from the Caribbean, Central and South America, and potentially from the continent of Africa from migrating to the United States, was defeated by the efforts and influence of Dr. Washington in a roughly ten-day period of heavy lobbying in 1915.

Even though Dr. Washington didn't always speak of Africa in the most positive way, you can hear in his tone a curiosity, a longing to know more, as he spoke of his mother and her lineage and what he did and, more important, did not, know:

> Of my ancestry I know almost nothing. In the slave quarters, and even later, I heard whispered conversations among the coloured people of the tortures which the slaves, including, no doubt, my ancestors on my mother's side, suffered in the middle passage of the slave ship while being conveyed from Africa to America. I have been unsuccessful in securing any information that would throw any accurate light upon the history of my family beyond my mother. She, I remember, had a half-brother and a half-sister. In the days of slavery not very much attention was given to family history and family records that is, black family records.[29]

I feel a similar longing to know, even as I pen this sentence. There is so much more that I want to know that I don't know and may never truly learn. We never had the chance to learn what our true roots were, how far, wide, and deep they spread.

Dr. Washington is rarely referred to as a doctor, but he was bestowed two honorary degrees, one from Harvard University in 1896 and one from Dartmouth College in 1901. In the Black community, the doctorate title is a recognition of seeing us and our presence. It grieves me that Dr. Washington was called boy or nigger in his lifetime, especially as he was growing up, much more than he is called doctor now, more than 100 years after his death.

Sawubona, Dr. Booker Taliaferro Washington.

Dr. W. E. B. Du Bois

A strong proponent of agriculture and the power and significance of land ownership was Dr. William Edward Burghardt (W. E. B.) Du Bois. Dr. Du Bois was painted as the antithesis and opponent of Dr. Booker T. Washington and his philosophy of vocational education as a viable path for Black families and communities in America. I was taught and have read that Dr. Du Bois promoted the Talented Tenth, a philosophy that promoted education in the arts and sciences of the most capable ten percent of the African American community to serve as an academic think tank and leadership cadre for the broader community, highlighted in his book *The Souls of Black Folk*, while Dr. Washington encouraged vocational trades. The two twentieth-century luminaries and cultural juggernauts had some philosophical differences, but there wasn't any lasting beef or discontent between them.

Dr. Washington served as a mentor and occasional adviser and sounding board rather than an adversary. On three occasions, Dr. Du Bois sought employment at Dr. Washington's Tuskegee Institute. Dr. Washington, however, didn't feel his students would benefit from the type of education style that Dr. Du Bois demonstrated. In my opinion, Dr. Washington erred in not seeing the social science and research value that Dr. Du Bois would have added to his institution and to Black academia. Can you imagine the powerhouse of an institution led by Dr. Booker T. Washington, with Dr. Carver and Dr. Du Bois as researchers on the staff?

Dr. Du Bois went on to be hired by the Bureau of Labor Statistics to do a statistical composite of African American life. His 1898 study *The Negroes of Farmville, Virginia* set the precedent for the bureau's publications on Black American studies and overall farm census projects throughout the early twentieth century.[30]

Despite early disagreements with Dr. Washington's viewpoints, Du Bois started to show his true leaves. In various writings he encouraged and supported Black agrarianism and Black landownership as a key source of self-sufficiency and community sustainability. In his role as a social scientist, he created some of the first infographics highlighting populations of Black farmers, the production of Black farmers, land gains, and losses—all of it a precursor to the work of the USDA Census of Agriculture. In 1921, he wrote about a small resort area in rural Northern Michigan called Idlewild.

(The movie *Idlewild*, featuring the recording artists OutKast, is not related.) Nature, and a connection to it, became an escape for the likes of Dr. Du Bois and other individuals and families who had left the South to live in northern and western cities.

Dr. Du Bois is more widely known as a highbrow intellectual and not a rural pragmatist. However, like many first leaves, he evolved and grew, shedding initial ideas and returning to an understanding of the needs of those who didn't share his ideologies. As a senior statesmen, after having public and private disagreements with Dr. Booker T. Washington and the Hon. Marcus Mosiah Garvey, Dr. Du Bois instituted some of their philosophies and had many of them implemented on the African continent with the 1950s and '60s waves of the African Spring movement. The first leaves of Dr. Du Bois's thinking morphed dramatically during his life, but supporting the Black and African farmer was a constant in his life and works. Sawubona Dr. W. E. B. Du Bois.

COCOA AND VANILLA

In the United States, nearly every house, school, gas station, and grocer is full of cocoa products. Every Valentine's Day and Halloween, chocolates are the sought-after and celebrated star. There's an amusement park dedicated to chocolate. To say it's just a confectionary dessert would undervalue the role of its culinary and economic value globally.

Cocoa (or cacao) is native to the Americas, where the Mayans, Aztecs, and Toltecs had developed its cultivation and use more than 3,000 years ago, before the Spanish arrived and got interested in this unfamiliar food. Literally translated as the "food of the gods," its Latin name *Theobroma cacao* signifies the divinity of this succulent seed in its multicolored pods. (See page 9 in the color insert.)

Africa became a site of cocoa production in the early to mid-1800s, and it was a Ghanaian blacksmith and innovator, Mr. Tetteh Quarshie, who founded the commercial industry in Ghana (then called the Gold Coast and colonized by the British). Mr. Quarshie is credited with bringing the first cocoa beans to what's now Ghana after a journey to the Spanish colony of Fernando Po (now Bioko in Equatorial Guinea).[31] In 1879, he successfully planted some cocoa seeds in the town of Mampong, part of the Eastern Region of Ghana, about 15 kilometers north of Aburi. Friends and relatives did the same, and the industry grew from there, with beans and cuttings also sent to Sierra Leone and Nigeria. Ghana began exporting cocoa in the 1890s, and for much of the twentieth century it was the world's biggest producer of the crop. Sawubona Mr. Tetteh Quarshie.

In 2025, Ghana was number two, behind Cote d'Ivoire, in cocoa-producing countries. Of the top six cocoa-producing countries, four are in West Africa.[32] To inherit a cocoa farm in Ghana is considered a great fortune.

Chocolate and vanilla are sometimes thought to be at war with each other—brown versus white, competing in some kind of confectionary arms race about which ice cream flavor people enjoy more. Historically, though, there is a long tradition of mixing the two flavors. Vanilla is another native crop to the Americas, derived from vanilla orchids, indigenous to Mesoamerica. The Totonac people had domesticated these orchids by the twelfth century, using vanilla as a flavoring, a fragrance, and a good luck charm. When the Aztecs conquered the Totonac in 1427, they noticed this intriguing substance and came up with the idea of mixing it into a cocoa-based drink they called *xocolatl*, "the bitter water,"—the forerunner of today's hot cocoa, and the source of our word *chocolate*. The brilliance and elevated taste palates of the Aztecs and Mayans led them to use vanilla to enhance chocolate. Sawubona Mayans, Toltecs, Olemecs, and Aztecs.

Like chocolate, vanilla caught the eye of the Spanish conquistadors—including the chief invader himself, Hernán Cortés. He is thought to have taken both chocolate and vanilla back to Europe, where they proceeded to become tightly woven into the food traditions of many nations.[33]

The Europeans couldn't figure out how to *grow* this plant, however; they had to import vanilla from the New World. It took them more than 300 years to realize that unless the vanilla flowers were pollinated by a specific insect—a bee found only in Mexico—they wouldn't produce bean pods, which is the plant part processed to derive vanilla flavoring.

The French had an idea (I wouldn't be surprised if it came from an enslaved or Africanized French person) that vanilla could be grown off the eastern coast of Africa, and in 1819 they shipped fruits to their colonies on the islands of Réunion and Mauritius. It would be another twenty-one years before any growers there realized a crop—and the person who made it happen was an enslaved African boy who wasn't even a teenager. His name was Mr. Edmond Albius, and he was a twelve-year-old genius. Sawubona Mr. Edmond Albius.

Young Edmond Albius. *Photo by Ambre Troizat / Wikimedia Commons.*

He was born in August 9, 1829, into slavery in Île de Bourbon (later called Réunion) and never knew his parents: His mother, Mélise, died during his birth, and his father, Pamphile, was enslaved on another plantation on the island. When he was very young he was sent to work on the Bellevue Plantation for a horticulturist named Ferrèol Bellier-Beaumont (who was also the plantation owner). This botanist had been trying to produce vanilla but had been able to keep only one vine alive, and in twenty years, it had never borne fruit. It's told that Bellier-Beaumont would have young Master Edmond accompany him around the plantation and share what he knew about horticulture with the young plant phenom. Young Edmond studied and mulled over what he had learned about hand-pollinating a watermelon plant, and he applied some of those principles and understandings to the mysterious and, at the time, nonflowering vanilla plant. He devised a technique of smashing the anther sac and stigma of a vanilla plant between his forefinger and thumb. This is known today as the marriage. If the marriage is successful, the base of the flower will swell within a few minutes. The swollen base then evolves into a thinner seed pod resembling a human finger, and in nine months, the vanilla pod is ready to harvest.

Plantation owners were invited to see young Edmond's technique firsthand, and he was taken around various plantations on the island to demonstrate this technique. But I find myself wondering: Did Mr. Albius learn from Bellier-Beaumont, or was it the other way around? How we craft these stories determines a lot about how we perceive the world. If Bellier-Beaumont is personified as a helpful teacher, then it's harder to also characterize him as a vile plantation owner. The unnatural stench of owning humans and all of the works from their brilliant minds, hands, genitalia, and wombs is overlooked. Bellier-Beaumont did do the pseudo-honorable action of emancipating Master Edmond in 1848, six months before the island territory made slavery illegal. However, that does not excuse the fact that the owner forced Master Edmond to work against his will for seven years after his industry-creating discovery. Mr. Edmond adopted the surname Albius, in an ode to the vanilla orchid that he worked with in his youth, a variation of *alba*, the Latin word for white. Bellier-Beaumont consistently wrote letters and vouched for Mr. Edmond Albius to be seen as the father of the vanilla industry in Réunion and be included in its written history for time and memoriam, as many sought to claim young Edmond's technique as their own. He also fought for Mr. Albius to be paid a yearly stipend by the state and the planters who handsomely profited from his innovation, because work was challenging for Mr. Albius to find. The only work available for newly freed enslaved Africans was low-paying manual labor jobs.

Slavery, as human rights advocate and entrepreneur Mr. Shawn Rochester explains in *The Black Tax*, was a 100 percent tax on your being. Your mind, body, soul, offspring, creative genius, botanical brilliance, ingenious engineering, and culinary cultivation was taxed 100 percent with impunity. Mr. Edmond Albius has influenced the production of wafers, ice creams, cakes, and candles around the world, and little is known of him, his name almost lost in the unkept log books of a French human trafficker. In 2004, through writing the book *Vanilla: Travels in Search of the Luscious Substance*, author Tim Ecott came across Mr. Albius's name in the Réunion archives, shining light on this lost and forgotten African botanical hero.

One of the earliest photographs in that part of the world, captured in 1847, was of a young Mr. Albius right before he was able to live life as a free man. Slavery was abolished in French colonies in 1848, and at age nineteen, with very little money to his name, Mr. Albius found himself trying to survive as a kitchen worker. He ended up doing some prison time and dying in poverty at age fifty-one. Meanwhile, several attempts were made to discredit him as the inventor of the vanilla pollination technique.[34] Today there's not a lot of information about him, other than the research done by Tim Ecott. Any imagery that accompanies stories about Mr. Edmond Albius portrays him as an adult, downplaying the fact that he

The red clay soil of Carter Farms has grown crops, animals, family, and legacy. *Top photo courtesy of USDA.*

Agriculture Intelligence flowed from the Shirley Plantation (*top*) to Freetown (*opposite, top*) through well-known chef and author Dr. Edna Lewis (*opposite, left; photo courtesy of Philip Audibert*), and it also flows through the next generation at Carter Farms (*opposite, right*).

ANGUS CATTLE
FREETOWN FARM
"FREETOWN" CIRCA 1868
JJ 38
EDNA LEWIS
(1916-2006)

Oil palm trees (*top*) and oil palm processing (*bottom*) in Ghana. *Photos by Marco Schmidt / Wikimedia Commons (top) and Uzabiaga / Wikimedia Commons (bottom).*

Farmer's market in Ghana; Michael and family in Ghana.

Gboma (*top*) is an African native plant, and horsenettle (*bottom*) is its North American lookalike. *Photos by Ubkoumbogny / Wikimedia Commons (top) and Nativeplants garden / Wikimedia Commons (bottom).*

PICTORIAL REVIEW

Ghana honors its agricultural heritage, displaying it on its currency (*top left*). *Photo by Bank of Ghana / Wikimedia Commons.* In the United States, you can still find denigrating "memorabilia" of the people who literally grew America (*top right; bottom*).

Indigo dye pit in Nigeria. *Photo by Aminucrus / Wikimedia Commons.*

Indigo and cotton crops tended by enslaved people helped to build the wealth and power of the American colonies. *Photos by Adeel Ahmed / iStock (left) and Michael Bass-Deschenes / Wikimedia Commons (right).*

Peanuts fruit underground (*left*). *Illustration by Franz Eugen Köhler / Wikimedia Commons.* Peanut crop in Georgia (*right*). *Photo by Jacqueline Nix / iStock.*

Coffee plant. *Photo by YinYang / iStock.*

Cocoa tree. *Photo by muendo / iStock.*

African native superfoods we grow at Carter Farms (*clockwise from top left*): taro, okra (*photo by Len Worthington / Wikimedia Commons*), sweet potato leaves (*photo by Kostka Martin / Wikimedia Commons*), and hibiscus (*photo courtesy of Volta Presentation Farm*).

More superfoods (*clockwise from top left*): purslane (*photo by Laval University / Wikimedia Commons*), amaranth, nkotomere, managu, and Nigerian spinach.

Teaching alongside Mr. Clif Slade (*top*). *Photo courtesy of Aramark*. Brother Renard "Azibo" Turner with Mr. Richard Yates and Mrs. Chinette Turner (*bottom*). *Photo courtesy of Barrett Self.*

Mr. Duron Chavis, Founder of Happily Natural Day, has manifested Sankofa Community Orchard, a 5-acre urban farm in Richmond, VA, that features amazing cultural murals, over 120 fruit trees and shrubs, produce, and a protective greenhouse. *Photos courtesy of Duron Chavis.*

Talking with visitors to Carter Farms.

With Mr. P. J. Haynie at his rice farm in Arkansas (*left*). *Photo courtesy of Mekhael Carter III*. Chef B. J. Dennis making biochar (*right*). *Photo courtesy of Jonathan Cooper*.

Mr. Julius Tillery (*left*) destigmatizing how we view and perceive cotton in African American communities. *Photo courtesy of Julius Tillery*. Mrs. Bonnetta Adeeb (*right*) is the founder and co-director of Ujamaa Cooperative Farming Alliance and President of STEAM ONWARD INC. *Photo courtesy of Bonnetta Adeeb*.

A field of cotton at Tillery Farms. *Photo courtesy of Ayanah George / Falani Spivey.*

The next generation of Carters being planted in the red clay, as we wait for them and others of their generation to germinate into our next growers, producers, farmers, and future.

made his discovery and displayed his agronomic genius as a teen. It's like we still can't believe a young Black male could have been smart enough to pollinate the vanilla orchid and set the foundation for one of the most iconic tastes in the world.

This child genius studied nature and the pollinators, without a classroom—all practical, firsthand experiences while literally slaving on the job. He was a spiritual listener in nature just like Dr. Carver. His understanding of stigmas and pistils, plant reproduction, and orchids was and is unmatched. I'm sure many of us have tried and failed in keeping an orchid alive, much less trying to keep it alive and make it produce seeds.

That's just one step in a painstaking process of growing and processing the pods. This is highly skilled labor and it's still done by hand, which explains why vanilla is the second most expensive spice in the world, behind saffron.[35] Imagine Bellier-Beaumont walking through his garden and noticing two fruits on that vanilla vine he had been pampering for twenty years, then learning that it was young Mr. Albius who had successfully pollinated them.[36] Within days the prepubescent wonder was being taken around the island to show his technique to other growers, providing technical assistance and PhD level training to other enslaved farmers in the region, like what Dr. Carver would offer some seventy-five years later to sharecropping farmers in Alabama. It turned out to be the birth of a thriving vanilla industry on Réunion, then nearby Madagascar, which is still the biggest producer of vanilla worldwide. The vanilla grown from this region, which includes Réunion, Seychelles, and Madagascar, are commonly marketed as Bourbon Vanilla (in tribute to Île de Bourbon).

I still haven't quite understood why vanilla—which is used in over 18,000 products—is often thought of as "white," or why it's associated with being ordinary, when its flavor is far from ordinary, and its pods, paste, and extract are darker than me. I've had the delight and blessing of seeing skin the color of vanilla beans, yet it's never described in that manner. The description and perception of vanilla in comparison to its actual look and flavor profile is as nuanced as eyes.

COFFEE

How about some coffee with your chocolate or vanilla dessert? The importance and popularity of coffee is undeniable. When you go to a coffee shop to order your beans, you might see the names of places where this crop is grown: Sumatra, Java, Nicaragua, Brazil, Honduras, Vietnam. And you might see Ethiopia on that list. It's the place where this whole worldwide coffee obsession started.

The legendary accounts of how humans discovered the effects of coffee include a ninth-century goatherder noticing his goats getting all ramped up while eating the berries of a certain bush. Another concerns a mystic in the thirteenth century observing very energetic birds feeding on those same red berries.[37] Both these stories take place in Ethiopia. Sawubona Ethiopia.

Coffee spread to Yemen through Somali merchants and then to Mecca, Medina, Cairo, and throughout the Ottoman Empire. It became an essential part of Islamic culture and made the jump to Europe in the sixteenth century. Everywhere it went, coffee was embraced and used to stimulate concentration and vitality.

Coffee is an African plant, discovered and shared with the world by Africans, that in the United States represents a $90 billion industry and is a totally ingrained part of American culture.[38] It's been popular here since the time of the American Revolution.

Coffee even makes a cameo appearance in the mid-nineteenth century miniseries called the American Civil War or, as I refer to it, the War of Southern Aggression.[39] When the war began, the United States was getting most of its coffee from Brazil, and that coffee would arrive through the port of New Orleans. When the Union forces blockaded New Orleans, coffee imports were slashed in half. This was a problem not only for the home front but for the Union army, which considered coffee a key ration for soldiers.

The solution turned out to be across the Atlantic in the young West African Republic of Liberia, ironically founded by free Africans from the United States who had been cast off by the efforts and fundraising of the American Colonization Society. The American Colonization Society was an organization devoted to repatriate free Blacks and some enslaved Blacks back to Africa, with very mixed motivations and intentions, including expulsion of free Africans who could inspire the enslaved to actively seek their freedom. The first ten presidents of Liberia were American born Blacks. Liberia's second president, Mr. Stephen Allen Benson, was born in Maryland. President Benson's free Black parents had moved their family to Liberia when Benson was a child. By the start of the Civil War he had become a prominent coffee farmer there. He had been shipping small amounts of coffee to the US since 1848, selling it to a Quaker abolitionist merchant in Philadelphia who was offering slavery-free goods.

The United States hadn't formally recognized Liberia as a nation at that time, or made official trade treaties with the young nation, because Southern states had stood in the way on the grounds that for a Black diplomat to visit Washington would be inappropriate. But once Southern states were claiming not to be part of the country anyway, Lincoln was free to recognize Liberia, which he did in 1862. (A diplomatic visit from an African delegation finally officially happened decades later, with the Liberian delegation arriving after the intervention and lobbying of Dr. Booker T. Washington, in 1907!)

This opened the doors for a quick-growing coffee trade between the two countries. It also supported the American Free Produce movement, that sought to support the purchase and procurement of products that were not produced by enslaved labor, which was the precursor for ethically based marketing movements like fair trade and organic. A Black merchant from Philadelphia, Mr. Edward Morris, visited Liberia to advise farmers on coffee growing, and they wasted no time in rising to the moment. Liberian coffee imports to the Union increased fast, and the fact that the Union army was able to maintain a steady supply of coffee for its troops boosted Union morale quite a bit. It's been observed through reviewing thousands of diaries and letters by Civil War combatants, that the word *coffee* was used more than *slavery*, *Lincoln*, and *war*. Sawubona President Benson.

President Stephen Allen Benson. *Photo by Rufus Anson / Wikimedia Commons.*

Meanwhile, Confederate troops were scraping by with coffee substitutes made from acorns and sweet potatoes, and were drinking hard liquor instead of getting caffeinated before battles. The coffee advantage was part of what won the war, and various truces and cooperations were created because

HONEY

No, Africans didn't invent honey; nature did. But the domestication of bees and the fermentation of honey began in Africa. Honey has been found in the tombs of various pyramids and edifices in Egypt and Nubia (Sudan region). Like the kola, honey held spiritual and cultural significance in many African cultures. Anywhere that people could make a container for it, honey was prevalent.

Honey was one of the first substances to be turned into value-added products. Honey wine, commonly known as mead, is one of the oldest known brewed or fermented beverages. The Abyssinians, known as Ethiopians today, stake a claim to being the home of *tej* (in Ahmaric), *mes* (in Tigrinya), or *daadhi* (in Oromo). As it was in China and other places, in Ethiopia this was a royal beverage, with those in lower caste relegated to *tella*, a beer composed of various grains—that, depending on the region, included sorghum, teff, barley, wheat, or corn—fermented with buckthorn leaves.

Symbolic of the bees and honey themselves, ancient Ethiopians had to pay a honey tax after collectively gathering honey from their vast highlands. Like female worker bees that collect pollen and bring it back to the hive to service the queen, Ethiopians serviced the kings and queens of Ethiopia with honey to make the royal tej. Honey taxes weren't eliminated until 1935, which preceded the dissolution of the monarchy of Ethiopia only by some thirty-three years. Tej was drunk by nobility

An ancient Egyptian image of a bee. *Photo by Keith Schengili-Roberts / Wikimedia Commons.*

in fine glassware or by the most elite in the sacred horn of a rhinoceros. The politics of mead was unique, and was carefully curated in the culture to serve not just nobility but also provisions for the military of the Abyssinian nation. Sawubona bees.

of coffee.[40] Union newspapers called it a "great loss" when Benson died in 1865. But the new Liberian industry he had jump-started kept going strong, and Liberian coffee became the preferred variety in America. It was better-tasting than other types and, as a crop, was more resistant to disease.

Americans—including Union soldiers who made it through the war years drinking coffee three times a day—developed more and more of a coffee habit. By 1885, the country was importing twice as much coffee as it had before the war, and the Unites States has been a caffeinated nation ever since.

KOLA

If we're not drinking coffee, we're drinking Coke. Well, there is no Coca-Cola, Pepsi-Cola, or RC Cola without the African kola nut. The story of Coke (the drink) usually starts with the Confederate soldier-turned-chemist Dr. John Pemberton. Pemberton, like many other aspiring beverage entrepreneurs, created many concoctions and formulations using various spices, roots, and herbs from around the world, claiming medicinal or performance effects. Kola nuts, like most African nuts, seeds, and crops, were introduced to European colonizers residing in Turtle Island by . . . you guessed it.

Kola has a bitter taste and contains a lot of caffeine, as well as theobromine, which also can be found in coffee, tea, and chocolate. It also contains kolanin, which is known to be a heart stimulant. In Ghana, where I was first introduced to kola as the original Jolt or five-hour (sometimes five-day) energy supplement, the kola tree is called *kyereye*, and the giant kola tree is known as *watapuo*. While I was working at a tomato factory in Wenchi, Ghana, the breadbasket of the country, our two security men, Mr. Bawaa and Mr. Yeboah, would be seen chewing on these nuts at about sundown. They explained that this nut gave them energy, using the universally African forearm symbol of strength, virility, and stamina.

I'd seen these uniquely shaped little nuts sold in markets and on the side of roads all over Ghana, but I was somewhat oblivious to the fact that they were one of the key ingredients in the most iconic beverages in the world, colas. When merchants told me the name of these nuts, it instantly clicked. And I'm sure it clicked with Pemberton while he was mixing these red, green, and brownish hard nuts with water or wine. His first beverage creation was Pemberton's French Wine Coca—doesn't that just roll off the tongue? It was a wine containing coca leaf, damiana herb, and kola nuts. He later removed the wine as the base of the drink and added a sugar syrup base, along with a few more herbs and flavors, to create a unique taste: the first Coca-Cola drink. The kola extract was part of the drink formula throughout the twentieth century, although it's no longer present in the popular beverage today.

Kola has been around West Africa for hundreds of years, and it is sacred among the Igbo people of Nigeria who use it in a multitude of ceremonies and events. Kola's legacy in Igbo land goes back centuries. There's an Igbo proverb: "Onye wetalu oji, wetalu ndu" ("He who brings kola-nut, brings life"). Life, energy, and spirituality are all contained symbolically in this iconic nut, which is also considered an aphrodisiac. A special wooden platter is reserved for the presentation of kola nuts exclusively. The chieftain, king, or elder says a prayer of longevity, peace, and protection along with the passing, breaking, and sharing of the kola nuts. In most traditional cultures, in public settings it generally isn't given to women. But women do consume it in their own ceremonies and events where men are not present.

The number of cotyledons (lobes or pieces) that make up a kola nut indicates its individual spiritual significance and who should or should not consume it. One-lobed kola (*oju obu/oji mmuo*) is reserved for the ancestors. Two-lobed kola (considered the bad omen kola) and seven-lobed kola (*oji asaa mmadu na asass mmuo*, the perfection kola) are not eaten by humans. The four-lobed kola (*oji udo na Ngozi*) symbolizes peace and blessing. It represents two males and two females, signifying completeness and the four Igboland market days of Eke, Orie, Afor, and Nkwo.

Interestingly, kola nuts were also a key ingredient in the transatlantic slave trade, being used to sweeten the otherwise unpalatable water carried on slave ships.[41] People would chew the nuts without swallowing them, and doing so made water taste sweeter and lessened hunger pangs.

This nut's impact on the world has been just as great to some cultures and nations as it has been to the Igbos and other West Africans. Good or bad, life would be different without having acccss to the world's most popular drink, a kola-inspired and many times kola-branded beverage. Sawubona kola nut.

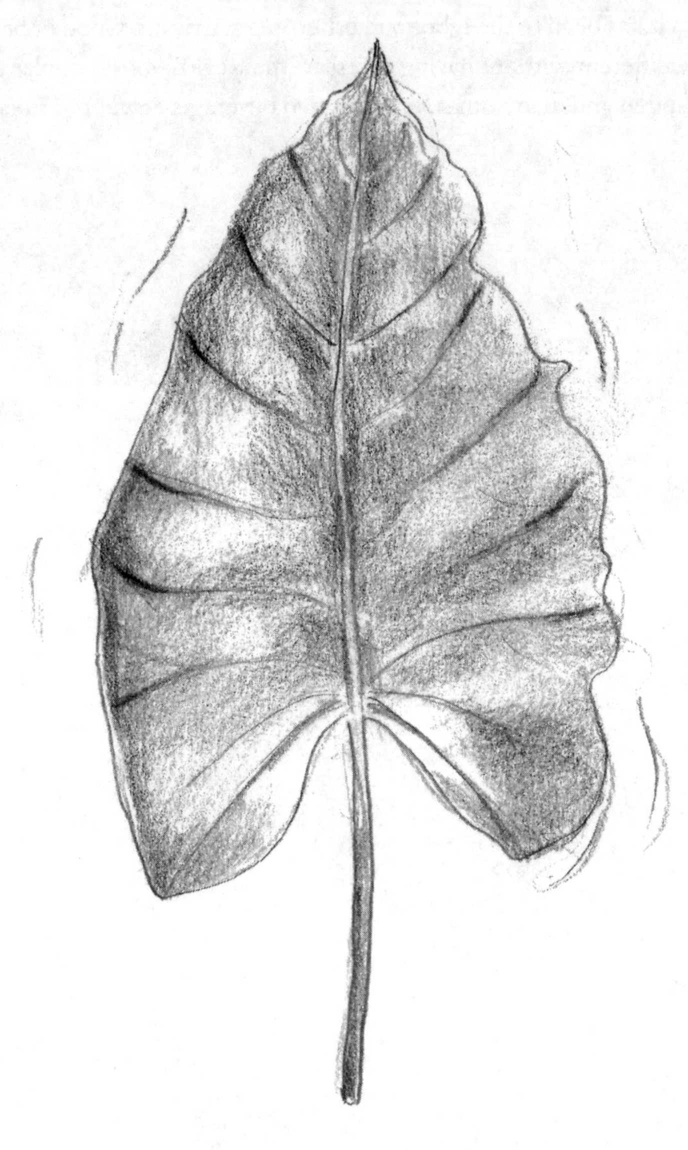

CHAPTER 6

TRUE LEAVES

THE NEXT WAVE

Even without drumbeats, banana leaves dance.

African Proverb

In July 2010, I took a trip to Ghana with my then wife, Eshmayah, and stepson, Nitsaviel. I had gone to West Africa twice before that, but Africa was new to her. My stepson was in ninth grade at the time, and we traveled along with his dance troupe, the Taratibu Youth Association (TYA). When we flew over Ghanaian airspace and Eshmayah saw the Ghanaian land for the first time, tears spiraled down her smiling cheeks, in a controlled release of long-sought satisfaction and fulfillment. And when she stepped out of the plane onto African soil, she started to cry, releasing emotions I'd never seen from her before, while she jumped up and down like a *Price Is Right* contestant who'd just won the final showcase.

Reactions like the one Eshmayah had to landing in Africa are a sign that this is something the spirit says it was wanting. A true leaf was being revealed through her raw initial reaction. At the time, we talked about moving there in four or five years, which would be after Nitsaviel finished high school. But just one year after that trip, in August 2011, we were talking about bumping up the schedule. Eshmayah was home in Maryland, pregnant, and I was back in Madina, Ghana, to do some reconnaissance. I was texting back and forth with her and was surprised when she said, "Oh my goodness, we're having an earthquake."

The epicenter of that quake was about twenty minutes from my grandmother's house and the family farm. Some of the foundation of her house cracked. Earthquakes are unusual in this area. At that time (or any time in my adulthood), I didn't like the direction this country was going in—social influences, the all-pervasive phones, and a culture of unaccountability. I didn't feel like this country would be the right place for my children to grow into their true selves. So Eshmayah and I became focused on getting out of America, and we decided to leave in May 2012.

It was not the easiest transplanting experience. There were numerous and quite unexpected trials and tribulations, including not having running water for two months, and my near-fall into an open sewer, but we persevered and stayed.

Our experiences and budget realities challenged the Costco, Price Club, Sam's Club wholesale mentality I'd gained while living in the United States. I'd go to the market and look at the price of pineapples: 25 cents apiece. Oh goodness, I'm going to buy ten of these, that's only $2.50. By the third day these sweet fruits were overripe, juicing themselves on the floor. I started to understand that Ghanaians shop daily, for the most part, because the food is so fresh. In the United States, we would shop only twice a week but that's because there are so many preservative-infused foods on offer. The Ghanaian markets and the market women elevated my diet and my consciousness around fresh whole foods, teaching me without saying a word. Sawubona market women!

I started to put a different value on the process of eating. Food preparation in Ghana is completely unlike how the packaged food in the United States is prepared, where you just put a meal into the oven or microwave and walk away while it cooks. Enchanted, I observed meals being made outside in a substantially sized, heavy aluminum or cast-iron pot—what many would term a cauldron—set over a coal pot (a metal frame that holds charcoal). I started to see why Sunday dinners were so important to certain families—the way the women and oftentimes men would be in the kitchen together, children playing, and then at dinnertime everybody would come together. Everybody had their role. The men made themselves available for going to the market, pounding *fufu*, which is a staple starch consisting of green plantains and either cassava or yam, or stirring corn dough for the dish known as *banku*. The stirring of banku is equivalent to stirring cement, the heat making the porridge thicker and thicker as you persistently stir, needing more and more strength and effort to move the twenty-four-inch wooden spoon. And the coal pot doesn't have a low or medium setting; it's just high. So there's an art and subsequent mastery to meal preparation, and you're not going to go away while you're cooking. You're definitely not going to watch TV or talk on the phone while you cook. The act of meal preparation requires complete focus, engagement, dedication, and love. We're all doing this together, which doesn't make it a chore but a family and communal responsibility, a part of a whole; it's what is done every evening. You fetch water and chop wood in the morning and start preparing the meal around two o'clock, and every ingredient is fresh with the exception of tinned tomatoes or tomato paste. "You're invited" is a phrase that visitors, neighbors, or passersby often hear after the meal is ready to eat and the consumption has begun.

Our first home base was in Cape Coast in the southern part of Ghana. It was relatively easy to keep up my American lifestyle there. There was a community of African Americans and a lot of African American tourists, and some of my friends ran hotels and other businesses there. The community we were part of had a vegan restaurant, and there were other vegan options around.

In our second year in Ghana, we moved north to the town of Wenchi because I received a contract to be an agriculture consultant at a tomato factory and farm referred to as Wenchi Can. It was started by President Osagyefo Dr. Kwame Nkrumah, who from 1957 to 1966 served as the first prime minister and president of Ghana. In his great genius he set up state industries all across the country, over fifty state institutions, many of which are still standing and in use. He was overthrown in 1966 by a CIA-led coup while he was traveling out of the country.

While making this factory and its 1,000-acre grounds my new workspace, I came across old metal machines that were made in Yugoslavia, a country that no longer existed by that name, having broken apart in a revolution in 1989. The lightbulbs were still the originals from 1963. They weren't as efficient, but they were still working, which made me appreciate the lack of planned obsolescence in the items sourced and purchased. These were "developing countries" trading quality and value with each other.

We were eight hours from Accra, the capital city, and I was the only American there. I stood out. I heard myself described as obroni a lot more frequently in the markets, and I realized my cultural training wheels had been taken off. I was in a place much more focused on living and surviving and not on tourism.

The house we stayed in was part of a little compound with goats that came into the yard every day. It was owned by Mr. Augustine, who was a *susu* collector or dealer. Susu is an African saving program in which the members collect money during the month, and at the end of each month a different person gets the money that was collected. It's a rotating system of saving and supporting. Whatever money you commit to putting in, the susu collector is expecting that every day. It's a constant collection, and this gentleman did it by riding around on a moped. There was nothing digital. In the evening, I would see him writing his notes in his logbook, and he received a percentage of the collection for doing his daily rounds. I'd heard of a susu before, but I never saw one run and function, especially with the administrative and social aspects. Mr. Augustine was steady and faithful in visiting his susu participants to collect their deposits, and the trust the participants had in him spoke volumes about his character. Sawubona Mr. Augustine.

I was beginning to understand and appreciate a broader sense of community and care. When my youngest son at the time, Yahir, was about two years old, an event transpired that assisted in opening my eyes to what community care can do for us. My children and I had (and still have) a tradition of saying goodbye to one another when I leave the house, and performing a special handshake. One day I was walking out of my house to go to the factory and my son came and gave me the handshake. But I didn't realize that afterward he started following me out to the road. Eshmayah realized he had disappeared and called me, upset—where'd he go? By contrast, I was a bit calm, cool, and collected in the face of panic, and my worry was tempered because ever since we arrived in Ghana, our neighbors always watched us and always knew who we were. And sure enough, after twenty or thirty minutes, one of the neighbors saw Yahir and brought him back home.

I had a similar experience with Nitsaviel, when one of our neighbors down the road came to my compound and said, "I want to let you know your son is hanging around with some questionable people, I want you to keep an eye on him." I was surprised, and I asked: "Who are you, how do you know my son?" And there were constantly knocks on the door as neighbors asked for help with school fees or medical problems. As Black Americans we were being watched and observed closely. It's not always a bad thing to be an obroni, and the realization that I was a Europeanized Westerner assisted me in seeing my true character outside of an American context. It was a first step into refining myself and my perceptions and attitudes around Africa and African people. This revelation became one of my surprising true leaves in Ghana, and I had to grow into it as I became more culturally acclimated to being an African.

THE TRUE LEAVES OF CONSERVATION

That year, 2014, I also started to be more inquisitive about nature itself. The plants I was learning about in Ghana opened my eyes to the relationships among people and plants and cultures—especially some of the nuances of the lesser-known crops.

At the market, we'd purchase unfamiliar produce—leaves of gboma (*Solanum macrocarpon*); Nigerian/Lagos spinach (*Celosia argentea*); alefe (*Amaranthus cruentus* and other varieties); nkotomere, also called cocoyam leaf or elephant ear (*Colocasia esculenta*); water leaf, also called boko boko or fame flower (*Talinum paniculatum*); and ayoyo or jute leaf (*Corchorus olitorius*). The names that farmers and merchants called alefe, water leaf, and jute leaf varied depending on what region of the country I was in. Nkotomere was and is still my favorite, served as a stewed green with a melon seed called egusi. When it's cooked right, I could eat it all day (and some days and nights I did). The green referred to simply as gboma (the g is silent), stood out at markets. As I explain in chapter 2, gboma greens are in the nightshade family. Thomas Jefferson and others in the planter class warned that the leaves of nightshades, and sometimes the fruits, were poisonous. And yet in Ghana, I met a whole population of people eating gboma leaves. When I traveled to Kenya in 2013, another nightshade leaf crop, managu, was served at almost every home-cooked meal. Managu (*Solanum nigrum*) is like their national dish, their collard greens or spinach, a nightshade plant that seemingly no one got sick or perished from eating.

Why do certain plants like nightshades affect African bodies differently than they do European bodies? Or is it just due to a greater understanding of the plant kingdom on the part of Africans? Agronomy in the United States has been based on European traditions and US nutrition standards on the body composition and biome of Europeans. One possibility is that crops traditionally grown in Europe are more favored toward the body composition of

European people. Each plant species has unique nutritional needs and nutrient content. Some are more nitrogen-centered, others more fungal-affected, and most others need balance of the two. For example, kale is a leafy plant that thrives in a nitrogen-rich soil, but it doesn't form relationships with mycorrhizal fungi. Two African crops—sweet potato greens and Nigerian spinach—contain some of the same nutrients as kale, but they are dependent more on the fungal network to supply their nutrient needs. Their content of manganese and some other micronutrients is higher than that of kale.

The difference between medicine and poison depends usually in the dosage. As we all know, or think we know, arsenic is poisonous. It's been featured in various whodunits, detective series, and murder mysteries as a pure poison. Arsenic is found in apple seeds, and many people avoid eating the seeds because of that. But in *Minerals for the Genetic Code*, agronomist Charles Walters posits that arsenic is a profound requirement for the living systems of humans, especially in pregnant mothers carrying male children. Expectant mothers' arsenic level increases to ten times its normal level; if it doesn't, it could cause spontaneous abortion of the child. Walters notes research studies showing that when arsenic is combined with another natural toxin, bryostatin, it is known to be effective against blood cancer, creating self-destructive actions in cancer cells. As for apples, the reality is that you would have to eat two cups of apple seeds in one sitting to consume enough arsenic to kill you. Eating the seeds in a single apple or two could actually have positive effects.[1]

In Ghana, considering these nuances of perception, practice, and our bodies' needs and intelligence aided me in seeing these leafy crops from a culturally and biologically different perspective. Collard greens and kale are brassicas that originate from Europe. Although Black people have made collards and kale our own, these plants don't grow naturally in the same climate Black people derived from. I refer to many of the traditional European vegetable crops that we grow in our gardens and farms as "spoiled" crops: Gardeners and farmers cater to them with drip or sprinkler irrigation and administer pesticides, fungicides, fertilizers, and/or pre- and post-emergent herbicides. The pH has to be perfectly right, so you may need to put down lime and other soil amendments, take out weeds, or grow them in greenhouses. They are spoiled rotten. But many of the African specialty crops we grow can survive in hotter weather and with far fewer resources.

I also began to realize that some of the African crops corresponded to plants I was familiar with from Virginia—like how the thorn-covered horse nettle is very similar to gboma. Amaranth has US cousins, too, such as pigweed, which also bears spikes and thorns to resist deer and wildlife pressure. Keen observations and curiosities can assist in our perception of weeds versus beneficial plants for human consumption or other beings.

I wanted to learn about every crop I saw. My wife was pregnant with our youngest son, and the midwife recommended she take iron-rich Jewels of Opar, or fame flower, sometimes referred to in parts of West Africa as water leaf or boko boko. Purslane grew wild, and we harvested it as our source of omega 3 fatty acids. As a preventative measure we took neem oil to keep malaria at bay. Malaria is transferred by mosquitoes that deposit parasites into the blood; as long as you're taking a substance that suppresses the parasites, you're OK. Neem oil makes the parasites deform when they reproduce—that's why it's used as an insect repellent. I took neem oil on a bimonthly basis for about three years and then when I stopped, I got malaria.

Working as an agriculturalist in Ghana, I found myself being humbled daily, as the realization dawned on me that I was not as smart as I thought I was. I started to appreciate and remember some of the things that were in my garden in Cape Coast, like a basil I didn't have any memory of planting. This basil was four feet tall and covered the front of a planter, while nothing else was growing well. I learned the Ghanaians had a much better developed understanding of the plants used for promoting health and vitality. There were so many herbs at the market for different ailments—tree barks, leaves, seeds, flowers, roots in bundles, bottles of bitters (alcohol-based tinctures) with herbs at the bottom. I became mesmerized by the Indigenous skill set of Ghanaian farmers and growers, skills that during my first two years of living there, I had hardly noticed. I also became greatly enamored with the art of living that I had been witnessing—how health, community, harmony with nature, emphasis on family connections, respect, and happiness are at the core of a well-lived life.

Everywhere I went, I asked questions about this plant, that bush, why those trees grew tall. I began to understand the conservation practices that had been in place for generations. Certain trees are called "no-man plant," as in no man planted them. I was told, "Those are the grandfather trees"; as in they had been there since their grandfathers' grandfathers' time. Sacred

SUSPITION OF POYSONING

There are stories of enslaved people poisoning slave owners, and I draw a correlation between these stories and the poisonous nature of nightshade leaves. The victims of kidnapping and human trafficking who arrived here from Africa understood these leaves were poisonous when not cooked well, and they would intentionally not cook them well in order to punish their enslavers.

An enslaved woman named Ms. Eve who lived right in my home county, Orange County, Virginia, was one of the first Africans in America found guilty of poisoning someone in the American colonies. I'm claiming her as an auntie, as well as claiming her innocence for seeking her freedom.

Auntie Eve was accused in 1745 of poisoning her enslaver, Peter Montague of Orange County, as he had died from being poisoned in December of that year. The charges said she had been "led and seduced by the instigation of the Devil . . . with force of arms and her malice forethought, feloniously and traitorously did mingle and poison milk . . . did give it to the said Peter Montague."[2] She pleaded not guilty but was convicted and sentenced. In January 1746, she was burned to death, and "the smoke of the burning of Eve was visible over a large extent of the country."[3] Sawubona Auntie Eve.

A decade earlier, in 1732, at James Madison's residence, Montpelier, three enslaved humans were charged with "Suspition of Poysoning" Madison's grandfather Ambrose Madison. Their names were Auntie Dido, Uncle Turk, and Uncle Pompey. Uncle Pompey was hanged on September 7, 1732, and Auntie Dido and Uncle Turk were whipped in Spotsylvania County.[4] Sawubona Auntie Dido, Uncle Turk, and Uncle Pompey.

Between 1740 and 1785, poisoning was the second most tried crime in Virginia after theft. The Virginia House of Burgesses passed the poisoning act, Virginia Act 38, in 1748, which prevented those who were enslaved from working with herbs or plants in a medicinal way. What's the difference between medicine and poison? It's all in the dosage and how it's given. I understand these actions as acts of resistance. The enslaved didn't give up; they were always thinking about how to get out of their ungodly situation or at least make those who benefited from this peculiar institution suffer. These prisoners of war and capitalism took control where they could, not through docility, but with strength and retaliation. All humans desire freedom, and many ancestors played the long game to get what they so desperately desired. My uncles and aunties were master physicians for their resistance against the tyranny of enslavement.

forests surrounded roadsides and were seen in the distance in hills and mountains, where loved ones were buried. In those places, soils weren't disturbed, trees weren't cut, and only a select few were permitted to enter. These ways and practices helped to limit deforestation in certain areas by making them sacred; it preserved the land both culturally and environmentally.

Dr. Frank Kwekucher Ackah, a professor of crop science at the University of Cape Coast in Ghana, told me about the importance of certain trees and plants to West African cultures—such as the oil palm, which is linked to spiritual and culinary traditions. The sap of the palm tree, with a consistency more like a sugary water than a sticky sap, can be used to prepare palm wine, which is used in libation rituals and in traditional marriage ceremonies, and in making a distilled liquor called akpeteshie. (Its distilling process is amazingly similar to the way people make moonshine in the States. I don't know who got it from whom—if enslaved Africans had been doing this process in their homelands, and then continued it when they were taken to North America?)

Professor Ackah also told me about the kapok tree (*Ceiba pentandra*), where gods are said to reside, and the bitter melon. "The bitter melon has a very important story," he said. "It's a spiritual plant. You see bitter melon in people's homes; it helps to keep away evil spirits. The chiefs use it to protect them against evil spirits or somebody trying to fight them spiritually." Yams are another significant crop. Many communities in Ghana, especially along the Tano River, celebrate yam festivals during which they present yams to the river gods to appease or thank them for the harvest season.

These traditions tie into behavior restrictions that help to preserve the resources—like in fishing communities in the south of Ghana, you aren't supposed to fish on Tuesdays. It's a sabbath of sorts—it allows for schools of fish to replenish themselves, which helps them stay healthy. In the central region of Ghana where I lived, Thursdays (Yawoada or Yaw Day, which means the "day of the earth") serves as a day of rest for the land, and on Thursdays people cannot go to their farms to work. This relates to one of the Ghanaian creation stories: the Earth was created on Thursday. Professor Ackah explained that in other regions Tuesday or Saturday might be the sacred day when you'd be more likely to meet the gods, so on those days you'd avoid the farm or the forest.

One Thursday, I ventured into the forest, ignorant of the sabbatical day, and what was meant to be a few minutes' walk turned into a few hours of worry. I quickly ascertained there was no one in the forest tending to their farms. I spent the next two to three hours trying to figure out how to escape the forest maze I found myself wandering in. No cell reception only added to my worry. I found my way home eventually, and later I learned about mythical creatures called *mmoatia* or "mysterious dwarfs"—forest dwellers that play tricks on humans. Everybody in Ghana believes in them. It's said they are two to three feet tall, covered in hair, with their feet turned backward; they grow only as tall as a mango bounces off the ground. They can take you with them to their lair, and what may seem to be a couple of hours could actually last weeks, months, or a couple years. Professor Ackah told me that it's believed these creatures gather around large trees; they "can help bless you with good harvest, but if you are a bad person they can punish you." I'm still waiting on the results from my forest visit. Sawubona Nana Ackah.

Nature-related mythical creatures figure in almost all cultures, and it was refreshing to hear cultural myths with an African background and relationship. Myths teach values, and these stories helped me learn about sacred forests and about conserving lands and resources in a way that resonates equally well with a 3-year-old and a 103-year-old. It impressed upon me how intrinsic nature is to Indigenous cultures around the world.

The principle of conservation was supported in many ways in Ghana, though it took me a while to realize it. For example, in a nearby village named Simiw, charcoal was a major product. Every day as I rode down to Simiw, men and women were faithfully chopping down trees. What appeared to be destructive was actually an environmentally responsible and ingenious practice of conservation and preservation. The cut branches were stacked in large piles and covered with soil, and a slow burn was started. They were making charcoal, which generally was completed after a few days of a continuous slow burn. The acacia trees that were being cut had been planted because of their regenerative capacity; they grow back every year to provide more charcoal for the following season. Farmers were encouraged to use the black soil from these charcoal-burning sites for enhancement of crops. This black soil is what regenerative farmers call biochar.

SACRED INTELLIGENCE AND SHARED EMOTIONS

Ghana opened my eyes to another world of life and compassion. Witnessing an observant mother hen reacting like a human mother would upon seeing someone taking her eggs forced me to humanize and empathize with other creatures. This hen wailed, cried, fussed, and argued with the gentleman—Mr. Bawaa, the security guard—for twenty minutes, begging that her legacy be returned. Waiting to begin an eight-hour overnight journey to Accra from Techiman, I watched live chickens and goats being packed into the bottom hold of the bus. This led me to think of how people sometimes describe slavery by saying, "They treated us like animals." I found myself wondering: Why are we allowed to treat animals like this?

The lifestyle in Ghana was closer to nature and also allowed the true nature of the human body to shine forth. I was mesmerized watching people fetch water. I marveled at bodies of all heights, shapes, and sizes carrying forty to fifty pounds of water in containers on their heads or in their hands. It was a feat I struggled to do myself, but all around me everyone seemed to have the grace and elegance of a runway model, exhibiting perfect posture to accentuate the divinity of the human form. Throughout the continent of Africa, the majority of people exude dignity and grace in every step.

Once while traveling from Benin into Togo on foot, I had to carry about sixty to seventy pounds of drills, hammers, and other tools. I was struggling, aware that it might be a thirty-minute walk, or even longer, to travel that distance and pass through the gauntlet of a border crossing. A somewhat elder sister/angel, shorter than five feet tall and likely about age seventy, asked me, "Can I take it for you?" She put my luggage on her head and said, "Let's go." I was impressed, but I began feeling less than the vibrant, fit, young man I thought I was as I had to exert quite a bit of energy to keep up with her through the queues of travelers that lined the border. She and I never shared names or greetings, but I can still picture her health and vitality, something I'd rarely seen in a person of her age in the United States.

My experiences in Africa also led me to rethink my basic ideas about agriculture. I realized that I'd been taught to judge success from the very perspective of European colonizers who loathed the idea of doing physical labor in the fields. I defined success as anything other than the hard work of agriculture. Watching elder men and women on their farms with their legs spread apart and bending over from the waist to work in the field or sweep was a master class in farm flexibility and endurance.

My experiences on various farms, and just living in my homes in Africa, deepened my appreciation for working with hand tools like hoes and brooms, which were always fitted with short handles. I don't recall working with many long-handled tools. Rather, I had to learn from those elders to bend over from the waist to prepare the land or sweep a floor. My experiences in Africa increased my awareness of how farming and agriculture is not just farm work; it's a farm *workout*. It involves strength training, plyometrics, stretching, resistance training, flexibility, low-impact training, and calisthenics, especially with manual and hand tool applications. Daily, I entertained a new idea—that perhaps working in agriculture could actually be the expression of my true leaves.

I learned other dimensions of the work and challenge of farming when I accepted an opportunity in Sogakope, in the Volta region of Ghana, at Volta Presentation Farms. This region is the home of the Ewe people, who are considered the fathers and mothers of *Vodún*, an African spiritual practice that is the precursor to vodun or voodoo in the Americas, most notably Haiti. I managed a staff of about twenty people, including farmers, administrators, security, and support staff. Again, I experienced my American privilege in full effect—my family had security, a driver, two cooks, a tutor, a laundry person, and still I was able to save half my income while earning more than my entire staff's wages combined.

Volta Presentation Farms had started one of Ghana's first organic community supported agriculture (CSA) models and delivery services. They served American diplomats and expats and Ghanaian dignitaries, delivering a weekly box of traditional American vegetables: broccoli, cauliflower, tomatoes, sweet peppers, cucumbers, carrots, and lettuce. I learned how hard it was to grow these crops in Africa during the dry season, so we created a pump system to acquire water year-round from the Volta River. The farm was a healthy, lush oasis of green vegetables, and every insect within a 200-mile radius apparently came and ravaged what we were growing. The cabbage and broccoli were especially beaten up by pests, and I had to troubleshoot on the fly. We ended up draping insect netting over the beds, but the pest damage took a toll on produce quality over the ensuing six to eight weeks.

Soon after my Volta experience, I went to Aburi to work with the foundation of Mrs. Rita Marley, Mr. Bob Marley's wife. The foundation staff and I assisted in revitalizing her operation, creating and diversifying market opportunities from the wealth of her land, including the local pineapples, starfruit, soursop, and oranges on the thirty acre property. Working with networks developed when starting an organic farmers market in Cape Coast in 2014, we found a local market, and the Rita Marley Foundation farm grew slowly but surely.

As I continued to notice the humanity and intelligence in both plants and people, the touch-me-not plant, *Mimosa pudica*, also called sensitive plant, kept catching my attention. Its leaves open and close in response to stress or a shock, including being touched. It grew all over Ghana. When my father visited, my son showed him a sensitive plant in the backyard in Cape Coast. He was mesmerized, and spent several minutes playing with the plant, touching it and watching the leaves close up.

Mimosa pudica also captivated French scientists in the 1700s. Jean-Baptiste Lamarck, considered to be the father of biology, was fascinated by this particular plant. In a way, sensitive plant is the "first leaf" of science: This plant with African origins that stimulated Lamarck's curiosity by its unusual characteristics aided in the construction of the science of biology.[5]

When I returned to Virginia from Ghana, I recognized sensitive plant as a weed, but my agriculture teacher father had never known this plant before. And after I finally embraced my own role as a teacher, my own true form, I would use *Mimosa pudica* to teach people about tropism. By developing and teaching the principles of Africulture, I'm trying to construct a new field of study and understanding just like Lamarck constructed biology. The intelligence of the sensitive plant connected me to Ghana and back to Virginia.

HOMECOMING

True leaves are the prime plant parts that can carry out photosynthesis, a key process for a plant to have a full and healthy life and existence. In a process akin to photosynthesis, I received information, intelligence, and rays of light from my experiences in Africa and turned them into fuel for my maturity and growth.

My parents visited us several times during our stay in Ghana. During their last visit, my father and I were on a walk with the children, and he asked me in great seriousness, "Where are all the starving people? I've been here three times and I see food everywhere." His remark made me realize the extent of the brainwashing Americans have absorbed about Africa and African agriculture, based on cultural staples from the 1980s and 1990s, like the "We Are the World" song and promotional campaign and the Christian Children's Fund commercials, where the actress Sally Struthers said things like, "For just the price of a cup of coffee each day, you can feed a child in Africa."

And so, while I was living in Ghana, I first began to formulate the concept of what would become the Africulture organization. Ghana doesn't have the same type of sophisticated modern infrastructure as the United States, but it has qualities that are more important for a community to succeed. For example, there is poverty and homelessness in Ghana, but I witnessed sincere care and compassion for the well-being of the homeless and those who may not have had much financially. Character and values were just as important in many cases as how much money one had, perhaps more important. I realized that the fullness of real life in Africa was very much unknown to people in the capitalistic "modern" world I had come from. Could some of the lessons I had learned in Ghana—about community, a culture of conservation, and African food plants—be communicated to people back home?

During that last visit, as I drove my father from the airport, he said to me: "You picked me up from the airport the last three times. I need to pick *you* up." He meant, You need to come back and see your family.

My younger self would have pushed back against a statement like that from my father. I would have said, "I'm my own man, I'm gonna do what I want." But I thought of staff members I'd worked with in Ghana who had quit their jobs to go home, simply because their father had told them they were needed. They were independent adults, but there was respect for the father's (parents') words. Learning from that, I listened to my father and planned a visit home in summer 2017.

The proverb, "When the student is ready, the teacher will appear," was on my mind as my departure from Ghana approached. Nature, the landscape, the culture, the people, and African history were all sitting me down and teaching me, shining their light upon my leaves and helping me grow in wisdom.

Back at the family farm for the first time in many years, I saw what I couldn't see before, and the land began speaking to me. I felt elated, and I was deeply curious to learn more about my family's history, my Hen Asem (which means "Our Story"). I began to think that my true leaves, the true inheritance for me, were in the United States after all, not in Africa, at least for that moment.

Walking around the farm, down at ground level, I noticed something else. "That looks like the sensitive plant," I thought. "And that looks like gboma." They were botanical brothers and sisters of sorts—it was as if they were modified African plants. I wondered whether there had been an evolution of certain species plants to fit different climatic conditions, just as African people had to adapt and modify themselves to survive in the cultural conditions here in America.

The spirit of *sawubona*—"I see you; I acknowledge your presence and your existence"—is born out of a relationship with the land, the soil and the plants, recognizing that to be seen is the first acknowledgment of respect. As I practice, I try to see the so-called undesirable plants on my farm and acknowledge their existence and need to exist. I'm not saying I don't weed, because I do; I weed a lot. But in weeding, I have a sincere curiosity about what or who I'm weeding out. I want to know about those plants, find out their taxonomic name and classification, their functions, their properties, how they benefit, and how they contribute in the broader ecosystem. They are a creation of the divine, and they have sprung up because of the conditions that I have created—whether it's by doing nothing, by mowing, by tillage, by covering soil with landscape fabric, applying manure or soil amendments, or setting up a greenhouse on top of the soil. I've learned

THE WORTHINESS OF WEEDS

Weeds are labeled as such by a person who doesn't desire their presence in a certain place. But in many instances, the weeds are supposed to be there, and the stranger, a colonial or gentrifying species, is the undesirable seed. Societies who don't understand or see weeds' value, who don't understand their worth because they don't fit into others' vision of what a utopian lawn, field, or pasture should be, will try to eradicate weeds at all costs. Some manually, grabbing by the throat and trying to uproot you or smothering you with a silage tarp. Some with mechanical means, utilizing tractors, plows, hoes, shovels, or garden trowels to try to fully gentrify their front lawn or their ball field. Others through chemical compounds designed to destroy your nervous system. Whether it be glyphosate or crack cocaine, it destroys the structure—the soil or soul—of a community.

So-called "weeds" are often unnoticed until they appear in a place they shouldn't be, just like those who grew early crops like rice and indigo that made this country and the agriculture industry what it is. We are rarely seen, acknowledged, or given meaningful credit for the building and maintaining of the American foundation. People of African descent are not weeds, but society has tried to RoundUp us, eliminate us, relocate us , or gentrify us when our presence is deemed undesirable.

much more from the native "weeds" that sprout on their own than I have from the crops whose seeds or seedlings I've planted. Africulture would eventually grow into an initiative with many branches—uplifting African crops and African innovators and knowledge-keepers who helped grow American agriculture and also helping Black farmers in the United States. But one of the first true leaves of Africulture was based on my experiences in Ghana, selling "American" vegetables to American expats. That planted a seed in me—the idea of creating a business to serve a niche market. Back in

Virginia, I would flip this idea around and sell native African crops to people from Africa now living in America.

I started by talking with a friend, Mr. Kevin Onyona, who owned two African restaurants called Swahili Village in the DC area. I inquired with Mr. Onyona about being a produce supplier for his restaurants. I asked him if I could grow some managu for him. He said he'd love to have it. As things turned out, I couldn't sustain production of managu, and I only ever sold him one batch, but I'd proved to myself that there might be a market in America for vegetables that are familiar to Africans.

I felt sure I could grow the crops, because from May to September the climate and temperature in Virginia are very similar to that of the rainy seasons in the nations of West Africa, where days range from about 70°F to the high 90s. The key thing was to choose crops with shorter growing cycles. We couldn't grow the African tree crops or plants that have growing cycles more than six to nine months long because of the cold winters in Virginia. But a lot of other crops do just fine—amaranth, Nigerian spinach, managu, jute leaf, okra, Scotch bonnet peppers, sorghum, millet, and cocoyam grow well within the cycles of Virginia's warm season. And they don't need a lot of resources—which made me think about their potential to be resilient crops in the face of climate change.

These crops are not like the spoiled brassicas, and so forth. They offer great nutritional value based upon their root structure: their roots grow deeper, allowing them to draw micro- and macronutrients and water from the soil during times of drought and harsh conditions. They can survive and thrive in moderate as well as unfavorable growing conditions in warmer climates. No one I have known or learned from has ever taught a plant how to grow, and I've certainly never taught a plant how to grow leaves, resist pests, or create fruit. The true gift of a plant lies in its inherent ability to carry out these activities and more without human involvement or instruction. Farmers and gardeners are managers, deciding when the seed should be placed in the ground and at what depth. And they can manage the growing conditions for the plant to fully achieve its purpose. The more management that is required to help a plant grow to maturity, the less natural that situation is. Crops, including the crops I came to know in Ghana, crops that have a strong capacity to survive, will thrive not because of you but despite you.

Many of these plants have been considered noxious. This is ethnobotanical racism, too. It wasn't just people that were discriminated against, it was African plants, as well. The National Academies Press has published a set of books called *Lost Crops of Africa*. The volume on grains discusses the unjustified perception that the native crops in Africa are inferior to the introduced grains such as maize, Asian rice, and wheat. While in Ghana, I'd hear marketing campaigns on the radio and television to encourage people to support Ghanaian rice, instead of basmati or jasmine rice from Asia. It was common to see bags of rice from Texas and Arkansas in markets throughout Ghana. It's generally perceived that the foreign rice is the better rice—like the African American adage that the white man's ice is colder than the Black man's ice.

The prejudices against African cereal crops are apparent in the ways they're commonly described—sorghum and millet, for example, are classified as "coarse grains," "minor crops," "famine foods," or "seed grains" suitable for animals only. Thomas Jefferson described the cowpea, known in America as the black-eyed pea, as a food for animals, not people. Or maybe it could be a food for certain kinds of people. "Many crops are scorned as fit only for consumption by the poor," *Lost Crops of Africa* states. "[But] the plants poor people grow are usually robust, productive, self-reliant, and useful."[6] There was a time in history when the English refused to eat potatoes because they were "Irish food." These are stigmas that reflect on both foods and farmers. Oftentimes, a crop grown in Africa is assumed to not be a good food; it's only a famine food.

Lost Crops of Africa discusses how European biases affected perceptions of native African crops, which in turn affected the resources—including the types of land—that were devoted to those crops. Native African grains have been called low yield, but that's in part because in times of modern agriculture, farmers weren't allowed to plant those grain crops in the best soils. And many have been labeled subsistence crops—crops grown to be eaten by the growers, not for sale on the market. As a result, there are no statistics on their production or cost. That lack of data stops people in charge of food aid or agricultural programs from considering them as potential market crops. Not all crops grown by farmers are marketable, especially those without a proven record of sales or demand from customers. The negative perception of Africa has resulted in branding of its crops in the same way as people from Africa have been branded as less than, not viable, lacking in value in the consumer market.

The Western ideal, which is a construct of the dominant culture, emphasizes counting and measuring. This counting and measuring construct has morphed into profit-and-loss-centered capitalism that doesn't and shouldn't apply to health and nutrition. If something isn't countable, it doesn't make sense to the Western mind, and thus it can't be exploited to make "cents." The factors that contribute to good nutrition and good health can't always be countable or commodified. They are generally part of a holistic lifestyle. Eating a diversity of foods, especially from different regions, can aid in supporting our health in ways science or math hasn't figured out how to quantify. It's a tough concept to grasp for some, especially when their education and ethos is based on a fundamental metric of counting and measuring. As I've worked to introduce various crops to gardeners, farmers, supply chains, restaurants, and grocery stores, the challenge of acceptance is generally the same at every level. I must answer the questions of who will buy this, and for how much? Is it a viable product, and how will consumers see the value in something they may not be familiar with?

Vegetable crops have been marginalized and discriminated against and undervalued, just like the people in the land they come from. In the New Testament, one writer tells how the ruler of Israel at the time spoke of the origin of the personage described as Jesus or Yeshua: "What good thing could come out of Nazareth?" The botanical equivalent of this, which I refer to as the Carl Linnaeus effect is: "What good thing could come out of Africa?" It's a loss for African Americans who could benefit nutritionally from these crops. All Americans and all people could benefit as well; everyone could use more biodiversity in their gut microbiome. Biodiversity is what makes us healthy and productive.

RESILIENT, NUTRITIOUS NATIVE AFRICAN CROPS

In my travels and experiences, I've tasted only the tip of the botanical and flora genius found in Africa. The following are a few of the African specialty crops I experienced in West Africa that should be explored, researched, grown, and eaten more in the West. These crops and many like them, could hold the key to continuing successful outdoor agriculture in the American South and other regions as temperatures steadily increase. They also serve as nutritional powerhouses that should be further researched, explored, eaten, and made into the ultimate marketing moniker, a superfood.

I've grown and promoted these crops through my work with Africulture—favorite crops that, to me, represent the true leaves of African places and people. They're all significant in the sense that they have proven they can survive and thrive through the tests of time in the climates and growing seasons of the United States, and the tests of discrimination in the marketplace and courts of public opinion. And many of them yield a nutritional value greater than their most familiar "American" counterparts and can be grown in most locales in the United States.

Nigerian Spinach

In spring and fall, the orange, purple, red, or yellow plumes of celosia flowers are part of the scene at garden centers. They are the ornamental version of Nigerian spinach. (See page 11 in the color insert.) This crop is better known as *sokoyokoto* in Nigeria, and the flower plumes sit atop the thin stems of this six- to eight-foot-tall giant of a plant.

Celosia includes about sixty different species, and both Nigerian spinach and plumed cockscomb (the annual ornamental that often has variegated leaves), are *Celosia argentea.* The variegated varieties aren't designed to be eaten, though—solid color green leaves are the preferred edible type, with some varieties having purple or dark red variations of the green leaves. The name celosia comes from Greek word *kelos* which means "burn"—the Greeks saw the plume of the flower as being like a flame. In the Yoruba language, *sokoyokoto* means "the vegetable that makes your husband's face rosy," a euphemism to explain its perceived sexual potency potential. In Malawi, the name used is *kaphikautesi*, which means "eaten by lazy ones." That name acknowledges that you don't have to do much to maintain and grow Nigerian spinach—just plant the seeds and wait. Direct seeding works well as long as the soil temperatures are warm enough. I encourage planting after Mother's Day weekend (the second weekend in May), in the continental United States. In warmer climatic zones, you can always plant earlier. Soil temperatures need to be about 70°F to 75°F for good germination. Keep in mind that soil temperature is different than the ambient temperature or weather; mid-May is about when soil temperatures are above or at 70°F in most parts of the country. Use a soil thermometer to check, if you're unsure. Transplants would work as well, but I've only ever direct seeded this crop.

Seeds germinate in seven to ten days, then this plant grows profusely. I could literally plant just one Nigerian spinach plant, and I think I would be set for the rest of my life. It self-seeds, so you start with one plant this season and thirty plants may well sprout the next season. It grows like a weed without demanding the tender loving care that other vegetables need. At the same time, it's not overly hard to kill; if you uproot the young volunteer plants, they're done. One summer I planted fifty Nigerian spinach plants in our greenhouse and grew them without irrigation. They passed the lack of resources test, reaching eight feet tall, without receiving any rain or irrigation.

Nigerian spinach seems naturally resistant to pests. I've grown it for several seasons and very little bothers its growth or quality of its leaf. I have to cope with potato beetles on the nightshades, cabbage moths on the brassicas, and white flies and mites on various crops on my farm, but nothing I've observed eats the Nigerian spinach, not even deer.

Nigerian spinach is a nutritionally complex leafy green. It's more nutritious from a trace mineral standpoint than kale; being very high in zinc, manganese, copper, and magnesium. Research studies indicate that zinc is a good mineral for reproductive health, and copper has been shown to reduce the incidence of graying of hair in some individuals. I'm not attributing my glowing mane of non-graying hair solely to consuming sokoyokoto, but it may be a small factor in my overall health and well-being.

As far as preparation, Nigerian spinach can be eaten raw or cooked; it tastes a little like kale or Swiss chard, but it's not as pungent. For the nutrition it provides, the aesthetic of it, and the pollinators it brings (including native bees), this is one of my favorite African crops.

Please note that this crop is not a traditional "spinach." Many European missionaries and colonialists and pillagers used the term *spinach* to refer to green leafy vegetables wherever they encountered them, rather than learning the name the local community and culture used to refer to the crop. If Popeye ate and gained incredible strength from eating the crop that Americans call spinach, then I equate Nigerian spinach to the plant with purple heart-shaped flowers that gave the Black Panther in Marvel Comics his strength and superpowers.

The plant is becoming a favorite among gardeners, and we are working to create markets so that farmers can profitably enjoy growing them, too.

One of our missions has been to market and promote sokoyokoto and other crops like it as an easy-to-grow African superfood. Seed suppliers including Southern Exposure Seed Exchange, Ujamaa Seeds, Truelove Seeds, and Sistah Seeds offer these seeds for sale. The latter three offer African seed collections, and Southern Exposure has a wide selection of heirloom seeds with African and Indigenous origins and history, as well. Sokoyokoto is an ideal candidate for community gardens, and as an ornamental as well as edible landscaping. As long as there is summer-like weather and a little moisture to stimulate germination, this crop will grow.

This plant can give anyone a green thumb and is a great conversation starter. Its height is impressive, with its flowers serving as fuchsia-and-purple torches that signify a successful growing season. Leaves can be harvested after the plant is about a foot tall. It's a pick and come again crop, so as the plants become taller, more leaves will be available to harvest. Just pull them off the stem and more will grow back in their place. The leaves do not become bitter when the plant bolts (goes to seed). Flowers form about sixty to seventy-five days into the season, with a radiant magenta purplish pink plume of flowers. I've seen varieties with yellow and green flowers, as well. As the flowers lose their color and start to turn beige, that's a sign that seeds are maturing and are ready to collect. If you don't desire to have a lot of sokoyokoto plant babies the following season, pinch the flowers before they fade to prevent the seeds from forming. Round, tiny black seeds are housed in the taupe flowers as the magenta color fades from the plume and the seeds fully mature. Some will fall to the ground while the majority will be protected from elements by the unique triangular shape of the flowers. The seeds can be harvested by cutting off the flowers and shaking them into a bag or container. Stored in a cool, dry place for the winter, seeds will be ready for planting next season or to share with fellow gardeners and farmers.

A wide assortment of bees, wasps, and other insects gravitate toward sokoyokoto flowers. It's a great pollinator plant that provides nectar for bumble bees, mason bees, great black wasps, and various beneficial insects and pollinators. Its flowers serve as a light both in the field and in the natural world, as a benefactor for many, giving much but needing very little, expressing the intelligence that sits within its seed. It makes my cheeks rosy every time I look at it, although probably not in the same way as the Yoruba meant it.

Nkotomere (Taro Leaf)

Besides sokoyokoto, taro (*Colocasia esculenta*) is my other most loved crop. Taro leaf, *nkotomere*, as it is referred to in the Twi language of Ghana, belongs to a family of tuberous vegetables, which Africans in Africa and the Caribbean oftentimes refer to as cocoyam; in Hawaii, it's called poi. A family of closely related and similar-looking tubers in the Araceae family includes eddo and malanga, as well. As with sokoyokoto (celosia), in the United States, taro leaf is commonly grown as an ornamental—the uniquely named elephant ear. They often have variegated white, green, pink, or purple leaves, but those aren't the ones that are good to eat. The round tubers—the botanical term for them is corms—are planted generally as deep as the corm is wide. But they don't have to be buried deeply, just covered with soil no more than two to four inches below the soil surface. (See page 11 in the color insert.)

When seeking out edible varieties, look for taro tubers at Asian or African markets. Some grocery stores with a diverse selection of fresh produce and vegetables may also sell taro tubers. Planting edible varieties will also provide edible leaves for consumption. When I came home for a visit from Ghana, I saw taro corms in a grocery store. I thought, "I know this crop; I'm going to take it home to my parents' house and plant it and see what happens." I planted the roots, and in July I returned to Ghana. In September, my family and I moved back to America permanently and I saw that the elephant ear had sprouted from the roots I had planted. After surviving the coldest winter I ever recalled (in 2018), it sprang up out the soil in June, proving its capacity to overwinter even in very cold conditions and be grown as a perennial crop.

Nkotomere is the most widely eaten and most commonly sold leafy green in Ghana, and it's found in every market. While in the United States, I noticed nkotomere leaf was not sold at any grocery stores, or at any Asian, African, or farmers markets.

Elephant ear prefers moist soil, but even if it doesn't have irrigation, it still grows pretty well. I planted a row of it in a high tunnel and the tubers spread and returned season after season.

The tuber is the plant part that's most commonly eaten around the world. Poi, for example, is the name of the taro tuber dish eaten in Hawaii, and it is sacred in Polynesian cultures. As a leafy green, taro has to be cooked well, stewed over high heat, because it's filled with calcium oxalate (a similar compound is found in rhubarb leaves). This chemical compound forms raphides, which are pointy, sharp crystals that irritate mucous membranes and skin that can cause itching, swelling of the throat, lips, mouth, or tongue, and can create complications in swallowing. In Ghana, the leaves are prepared with palm oil and egusi (seeds of various cucurbit melon species). It's awesomely good. I haven't grown the egusi, but it can be acquired from African or Asian markets and grocery stores, both as whole seeds or in ground form.

The corms can be planted year-round; however, they will not germinate until soil temperatures reach around 75°F. You can expect leaf production starting between early May and mid-June, depending on the hardiness zone. Harvesting is simple, but it requires a little effort to pull the leaves off of the top of the stem, being careful not to uproot the whole plant. To avoid your hands itching, I recommend wearing gloves when harvesting. The leaves grow back every seven to ten days throughout the season. The younger, smaller leaves, between three inches and one foot long, are the ideal size to harvest. The stalks are used in various Asian cultures in stews and soups.

Amaranth

Amaranth is somewhat well-known as a grain crop, but amaranth greens are also edible. The grain can be turned into tortillas and breads, and the leaves are good for eating at any stage of growth. This crop has an image problem, though. You may have heard of the common garden weed called pigweed—it's a species of amaranth that certain individuals felt was food fit only for pigs.

There are approximately seventy-five species of *Amaranthus*, with species endemic (growing only in a specifically designated geographic location) to every continent except Antarctica.

Amaranth has carried a stigma over the centuries because of its origin and the skin color of those who mainly consume it. "Amaranths are a poor people's resource, and the plants are dismissed as 'lowly' and ignored. . . . Few species of vegetables are so looked down upon," reads a vignette from *Lost Crops of Africa* volume II.[7] In the Caribbean, you might be told you're not worth an amaranth, meaning you're someone who has no value and is worthless. But amaranths were domesticated in both the Old and New Worlds, and I know it as a prolific, self-reliant plant that germinates in as little as seven days after sowing seeds, at which point you can eat the young sprouts as nutritious microgreens or wait a few weeks longer to begin harvesting the true leaves. Known in the Nahuatl language of the Aztec Empire as *huauhtli*, it was used ceremoniously to honor Huitzilopochtli, the Aztec god of war, and it maintains a strong relevance in food cultures throughout North, Central, and South America.

The amaranth leaf ranges in size from that of a toddler's handspan to an adult's hand. (See page 11 in the color insert.) It doesn't require much care after planting. Leaf-chewing insects such as leaf miners, moths, caterpillars, and grasshoppers, among others, seem to enjoy dining on amaranth throughout the growing season. Leaves can be harvested continually, and unlike brassicas, lettuces, and spinach, amaranth leaves don't turn bitter when the plant produces a flowerhead. Its ability to thrive on its own seems to me a phenomenal gift.

I would buy amaranth greens at the Techiman market near Wenchi, and I became more familiar with it when I returned to the United States. In the US, I noticed cans of callaloo on the shelf at the grocery store or African market, but fresh leaves were not available. Callaloo is a famed dish from the Caribbean that is best served with fresh green leaves complemented by the spices of the islands. (This triggered a memory of an episode of *The Cosby Show* that I'd seen probably twenty-five years in the past. Dr. Cliff Huxtable and a Caribbean friend were making a pot of callaloo and the doctor kept trying to taste what was in the pot, and his friend hit him with a spoon.)

Amaranth grains are considered a source of complete proteins, with double the amount of protein found in rice and corn. The leaves, too, provide more protein than spinach, along with twice as much vitamin C, and more calcium and iron than spinach or chard.[8] My appreciation for this ancient plant continues to grow as I learn more about its nutritional and cultural value.

Amaranth grows phenomenally on Carter Farms, just like Nigerian spinach, and if I allow plants to go to seed, the crop self-sows into the next season. To prevent plants from dropping hundreds or thousands of seeds, cut off the flowers as they start to develop, before they have a chance to form seeds. Take note that the flowers on some varieties of amaranth stay the same color through seed formation and seed maturation.

If you want to save seeds for planting the following year, you'll need to contain the flower heads in small paper bags. As the seeds form, tie a bag tightly closed over each flower head you want to save seeds from. At the end of the season, cut off the stem below the flower/seed heads.

In my experiences, amaranth has a high tolerance for less than optimal soil conditions and can withstand periods of drought up to a few weeks. Like some of the other Africulture crops, amaranth could be a resilient crop choice in the face of hotter, longer summers and other challenges arising from climate change.

Okra

Okra is one of those plants that people either love or hate. Okra is related to coffee, cocoa, and hibiscus. You can see that relation to hibiscus when you look at the showy creamy yellow flowers with scarlet centers. Everything about this plant is beautiful—the five-petaled flowers, the five-fingered leaves, the five-sided slender pods full of chunky round seeds. (See page 10 in the color insert.)

Okra is native to Africa, specifically East Africa. Its West African Igbo name, *ọ́kwụ̀rụ̀*, is the source of our English name okra. Some Ghanaians refer to this crop as *okro*, or as *nkruma*. It came to the Americas on the sea vessels carrying kidnapped humans and became a widespread iconic crop throughout the American South.

One of the original names for the plant is gombo, from an Umbundu word (*ochinggõmbo*) or a Kimbundu word (*kingombo*). When grandparents, parents, uncles, and aunts say, "I'm going to make gumbo," that means they will be preparing their recipe for okra stew. The languages and culinary traditions of the captive Africans branded as slaves weren't respected by white people, and neither were the names of their foods. The fact that Africans in America called okra by the traditional name of gumbo for hundreds of years, yet white people failed to recognize that gumbo was the name of the plant and not a dish made from the plant, speaks to the gross disrespect of language and culture that sub-Saharan Africa faced for close to four centuries.

Like other Africulture vegetables, one of the big advantages to okra is how profusely it grows. Many also consider it easy to grow, and by easy to grow, I'm simply saying, plant the seed and walk away. As long as the soil is warm enough (soil temperatures above 70°F) and there's a little moisture to start the germination process, it will grow well. Like most of the crops I enjoy growing, okra thrives in hot weather. After Mother's Day is the best time to plant okra, either by direct seeding or transplanting seedlings. Harvest when the okra pods get about two inches long or whatever the recommendation is for the particular variety you are growing. To save the seed of the okra, leave the pods on the plant and harvest when the pod turns brown in color.

There is a growing interest and curiosity about eating okra leaves—they are mild in flavor, edible raw or cooked, and, like the pods, can be used to thicken soups and stews. Even the flowers can be eaten, and the seeds can be used as a coffee substitute, though they don't have the caffeine.

In the United States, aside from certain traditional Southern foods, we've generally kept okra out of our diets. Many people say they love gumbo but hate okra. OK, which is it? Okra gets a bad rap because it's mucilaginous—when cooked or sitting in water, the pods and the leaves exude a mucus or slime (or some people would say snot). But I don't disrespect okra like that. This is a plant that helps to lower blood sugar, reduces high blood pressure, and has benefits for female and male reproductive health. Women in Africa drink okra water—they'll cut up okra pods and put them in water and drink that slimy juice. You could also put okra water into a smoothie and blend it up with strawberries and blueberries. Okra leaves, especially when younger, contain the same mucilaginous attributes as the pods.

Our palates in America generally are not used to the slimy texture of okra, but the health benefits are something we probably don't appreciate as we should. Whether you eat okra raw, fried, candied, or pickled—eat it.

Managu

Managu is a plentiful, productive, and nutritious leafy nightshade crop in East Africa. Ever since I first ate managu in Kenya, it stuck in my memory as something I wanted to grow one day. I was intrigued that managu was a nightshade vegetable (another name for this crop is African nightshade) that people in many places in Africa ate all the time. Years later, while growing managu for the Swahili Village restaurant, I started to see the benefits of growing crops for niche African markets in this country.

Managu is similar to huckleberry in the way the fruit grows right at the nodes (the point where one stem joins another); the black berries can be used like blueberries. (See page 11 in the color insert.) It does have some insect pressures from pests such as thrips, grasshoppers, and leaf miners, and certain types of moths place tiny holes in the leaves. Like peppers or eggplants, managu plants have a thin but hollow stem; it grows about three feet tall and bears three- to four-inch droopy leaves. Managu self-seeds but not as abundantly as Nigerian spinach.

In Kenya, only the young, tender leaves are used since the older leaves are more bitter, and I'd recommend the same harvesting practice wherever it's grown.

Sweet Potato Greens

Traditionally on farms in the United States, the leafy tops of sweet potatoes are treated as rabbit, goat, or pig food; farmers have historically fed the greens to their livestock after they had harvested the tubers. But in certain Asian and African cultures, the leaves are eaten quite frequently, sometimes more commonly than the tubers. (See page 10 in the color insert.) I grew familiar with potato greens through my acquaintances and friendships with Liberians. As the ill-advised American I am at times, I'd think to myself, "You can't eat potato leaves; they're nightshades." However, I came to understand that sweet potato greens were the greens being referenced in our culinary and cultural conversations.

Vitamin A is just as significant in the leaves as it is in the flesh of the tubers that my grandmother Lucille used to make slices of heaven, better known as sweet potato pie. Sweet potato leaves grow plentifully and profusely, and they can be both ornamental and helpful to suppress weeds. Planting sweet potatoes can be done one of two ways: You can grow them straight from planting a tuber in the ground or you can plant the long stems that sprout from a sweet potato tuber. These stems are called slips. Slips do best in sandy soils, as clay soils can limit the overall size of the sweet potato, and are planted in various ways. I use the trusted method that my uncle Clif Slade uses, which is to lay the slips in a growing bed about three feet apart, and press the root end of the slip down into the soil with a bottom-notched stick or pole so that the roots end up one inch to two inches below the soil surface. Most growers do not bother to irrigate their sweet potatoes, but rather allow the summer rains to assist them to grow. Uncle Clif also recommends planting around Memorial Day weekend. (There's more about Uncle Clif's sweet potatoes in "Mr. Clif Slade" on page 247.)

If you harvest the individual leaves, picking them off the vine, they'll cut and come again—the tubers grow hundreds of leaves per plant, and if you cut them, the leaves may grow back fuller. This crop produces leaves so prolifically, each slip grows hundreds of leaves, that it's usually no problem if deer show up to feed on them. And insects don't chew on the leaves much, which is always a plus. However voles, moles, sweet potato weevils, and groundhogs can be a problem, feasting on your tubers underground.

I've promoted this as a good crop for some of the farmers in my network to grow and market because sweet potato leaves are plentiful in the fields and have very little representation in the marketplace. The greens are a value-add that you can harvest without hurting your sweet potato tuber yields. Dr. Carver wrote several Carver bulletins related to the sweet potato, promoting it as a sellable and marketable crop. He encouraged eating the sweet potato leaves as early as 1910.

Even sweet potato stems might be a potentially saleable, edible product. The stems can be very chewy and fibrous, but they can become tender if peeled, removing the thin outer skin like you would peel the string from a string bean, and blanched in water for one to 10 minutes. They can then be used fresh in stir-fries or soups.

Sweet potato leaves are literally a food out of this world. NASA, which has researched the sweet potato plant with a specific interest in its leafy greens, first sent a sweet potato plant into space in 1999, on the space shuttle *Columbia*. The plant has proven to be microgravity tolerant, and provides astronauts and space tourists with a high-quality, nutrient-dense food source. As it so happens, this dense leafy green grows in the same manner, quantity, and resilience in space as it does on Earth.

Jute Leaves

Jute (*Corchorus olitorius*) is a highly useful plant native to Africa and other tropical and subtropical environments on several continents. Jute, known as *ewedu* in some cultures, has been a staple in West Africa for centuries. It also has a long history of providing fiber for baskets, clothing, and rope.

In early America, jute was discussed as a potential source of fibers for ropes and cordage, but that industry never took off. Despite its usefulness, on American farms jute essentially became a weed. (It's usually called velvetleaf now.)

Jute is best known as a fibrous plant used for making a coarse fabric for burlap bags, but it is very much an edible plant. In Arab nations it's referred to as *molokhia*, and it's often served as a soup or stew. But jute has never been a popular food in America because of its mucilaginous texture. Like its cousin okra, it exudes the same type of gelatinous goo when cooked, and because of this it unfortunately was robbed of the opportunity of joining the American leafy vegetable superfood lexicon. However, it's still a potent superfood: a nutritional powerhouse that can help to regulate high blood pressure, reduce sugar levels, and aid both male and female reproductive functions.

The blueish green seeds are uniquely shaped, not round, but almost like an abstract polygon. It is a tall, spindly plant; its seeds can be planted from seven inches to about twenty inches apart. The seeds are small; cover them just lightly with soil. I use a stick to make a shallow furrow and drop in the seeds, using the stick to push soil over the seeds. After being watered manually or by rains, the seeds will begin to germinate in seven to fourteen days. Planting should occur after the second weekend of May in most parts of the United States, or when soil temperatures are above 75°F. The vibrant green leaves are harvested by tugging them gently from the stem. They can be harvested throughout the plant's growth cycle. Yellow flowers develop into brown seed pods about two to three inches long. Gently break open the seed pod between your fingers to reveal the blueish green, brown, or black seeds that are ready for planting for the next year.

Hibiscus

Hibiscus (*Hibiscus sabdariffa*) tea is one I often drank during my five year journey in and numerous visits to West Africa. The deep magenta-maroon flowered plant has a host of names, including *sobolo*, *bissap*, *sorrel*, and *roselle*. Every region has a different name for it. In Mexico, hibiscus tea is called *agua de Jamaica*, and throughout other parts of Africa it's *tsobo*, *zobo*, *wonjo*, *dabileni*, *folere*, and *karkade*. Mixed with a sweetener, pineapple juice, ginger, and spices, the tea has phenomenal flavor. It's also loaded with health and nutritional benefits: It reduces hypertension, acts as a diuretic, assists in regulating blood pressure, and according to some research, lowers the risk of deep vein thrombosis.[9] The dark magenta calyx (the outside covering of the roselle flower) is cut off the round white ball of the flower and dried. The seeds are inside that white ball.

Hibiscus seeds are black, and each one is about the size of a pearl. These can be direct seeded or transplanted, but I'd encourage transplanting. Start the seeds in April indoors and transplant the young plants outdoors on

Mother's Day weekend in mid-May. I encourage starting the seeds inside as early as possible, because it's not a fast a process. Start to finish, it has taken me five months of tending the plants, waiting for them to flower, and then harvesting the precious calyx to obtain hibiscus seeds to make tea. Plant the seedlings twenty-four to thirty-six inches apart, as they can grow up to be quite bushy. The deep magenta hues contrasted with the rich dark green of the leaves makes this a potential edible landscape plant, as well. (See page 10 in the color insert.) The flowers are floral works of art, a divine artist's rendition of white abundance with a distinct, pronounced yellow or white stigma, which will attract numerous pollinators and onlookers.

In some nations, young hibiscus leaves are also eaten. They are known as *sawa sawa*, indicating the sour or bitter nature of the young leaves. The Krio of Sierra Leone make their sawa sawa with egusi seeds, and it's a treat I hope to enjoy one day.

FARMALL

CHAPTER 7

PESTILENCE AND RESISTANCE

FROM PATHOGENS TO THE PIGFORD SUIT

The problem of the twentieth century
is the problem of the color line.
Dr. W. E. B. Du Bois

America . . . changes all the time,
without ever changing at all.
Dr. James Baldwin

As Dr. W. E. B. Du Bois stated in 1903, the problem of the twentieth century is the problem of the color line. Having lived through the end of the twentieth century into the twenty-first century, I have seen that this problem still hasn't properly been addressed or resolved in the broader society of the Western world.

It's one thing to face systemic racism and the predictable problems of living in a society that stacks the rules against you. But there are also unforeseeable events that descend like a pestilence—sudden, traumatic, impossible to explain.

Even a plant growing beautifully in the best soil, given every opportunity to thrive, can be injured, sickened, or killed by a random event or unexpected development: a mower blade that comes too close or an invasive pathogen. The same is true for people. But the truth is, most forms of pestilence aren't entirely random. Most, to some degree, can be foreseen. In agriculture, soil health manifests in the above-ground reality of the plants and how well they're able to fight back against pathogens and pests—healthier plants, from the roots up, are harder to kill. If soil is depleted, a plant has fewer defenses and can be targeted by pests that detect the plant's weakness. The same is true for plants that are removed from their community of other plants and fungal networks. Such plants live in a weakened state until they fully adapt to their new environment.

People, too, become more vulnerable to problems if their roots aren't healthy or don't have rich soil to nurture their growth. Remove a population of people from their native continent and enslave them in another land across the ocean, and you compromise their health and survival. Put up a series of legal and social barriers to the economic and political advancement of their freed descendants, and you guarantee that those people will be more vulnerable to all kinds of problems.

As discussed in earlier chapters, the mass extinction of Black farmers in America today is the result of centuries of destructive, targeted, and predatory policies. If you think of Black farmers as a crop, that crop was uprooted repeatedly, denied nutrients and water, planted in the worst possible soil, and subjected to pestilence. It's no wonder that, as a category, Black farmers have become so scarce.

But the other part of the story is about resilience and resistance. I don't like to use the word *resilience* often, because it's both a blessing and a curse. Resilience is a blessing because it allows you to keep going, but in some cases it also means you will continue to endure perpetually worse treatment because of your innate and/or genetic ability to endure, adjust, and continue to withstand. We can see this in our gardens and on our farms. Some individual plants are better than others at fighting off cabbage worms, aphids, thrips, potato beetles, and other pestilences. In a patch full of plants getting chewed up, there's often one standing out with hardly a mark on it.

Throughout American history individual acts of resistance—and collective ones, too—have kept Black families and communities, many of which are generally made up of Black farmers, alive and moving forward. Resistance is necessary, as part of a growth process, part of the life cycle, but it shouldn't define a species or a people and be the entire life cycle of that species. When too much time and energy is spent on resisting, it can limit or stifle creative genius. The Western world has truly stifled its own growth by cutting off beautiful, broad noses to spite smooth flawless dark-complexioned faces. The amount of genius and excellence that has spawned from being resistant and resilient and resolute in action is phenomenal, but how much richer would the world be if barriers toward our freedom, expression, unity, and greatness hadn't been blocking our paths for the last five centuries?

DIVISION AS PESTILENCE

The need for Black resistance and resilience began with the first interactions between Portuguese raiders and the inhabitants of the Canary Islands in the early 1400s. A Portuguese-born prince known as Henry the Navigator, or as I call him, Henry the Exterminator, was searching for gold as he cautiously tried to "explore" the West African coast. He became a living, recurring nightmare for the citizens of the Madeira archipelago, located about 500 miles from Portugal. His lust for wealth, and his sadistic alchemist experiment of trying to convert free humans into black gold, currency, and labor for Iberian exploits, spread to the Canary Islands and farther down the African coast. Mr. Howard French's phenomenal book, *Born in Blackness*, goes into meticulous detail on the pre-Columbus exploits of the monarchs who orchestrated and financed the early Age of Exploration. Mr. French's ancestors and my own may have worked alongside one another, as both our ancestries run through Orange County, Virginia, and specifically the area in and around Montpelier, the family plantation of president James Madison.

The opportunity in Africa seemed so ripe that the unofficial governing body of most of Europe—the Vatican—was about to establish its own laws (papal bulls and papal patents) to usher in an age of exploitation for their free, autonomous, culturally fulfilled, spiritually satisfied, happy, unsuspecting victims. Pope Eugene IV took some actions to "protect" the Canary Islanders, apparently from the same people seeking permissions, blessings, and support from the Roman Catholic Church. His 1434 papal bull, *Creator Omnium*, for example, condemned enslavement and demanded emancipation. His 1435 *Sicut Dudum* prevented the Spanish or Portuguese raiders from slave raiding Christian converts on the Canary Islands.

The next pope, Nicholas V, made the exploitation of African lands business as usual for the empire of the Roman Catholic Church. Pope Nicholas V's 1452 papal bull, *Dum Diversas*, authorized the Portuguese king, Afonso V, to conquer and enslave "Saracens (Muslims) and pagans" in Africa. In

1454, Nicholas V authorized Afonso V to build churches in all conquests and granted Portugal dominion over all the seas from India to Africa. In 1455, with the papal bull *Romanus Pontifex*, Nicholas V concluded his reign of authoritarian terror: he granted Portugal permission to hold a monopoly on trade with Africa and permitted the enslavement of natives of Africa.

The influence of the papacy crept into everyday European life; in 1517, Spain established the "Asiento de Negros" or Negro loan and short-term debt contracts between the Spanish government and exploitive Portuguese and Genoese merchants for monopoly contracts to provide enslaved Africans as workers for their new project colonies. Spain had a sizable population of Moors, a darker-complexioned, majority-Muslim population with African roots who resided in the Andalusian peninsula and throughout Europe since the eighth century, and made it illegal for Negroes to pursue higher education. Spain enacted one of the first Black tax laws, which forced a special tribute to the government for being an educated and free Moor, or Black person, in Spain. This would be a precursor to a similar tax or "tribute" required by free Black males in the American South known as capital taxes. These types of accounts set the standard for how free, productive, talented, minding-their-own-business people of melanin would be seen. These laws authorized the conquest of Africans by Europeans, even though Africans had no idea they were at war, or in a race, with those Europeans.

Why did the Pope think he had the authority to decide the fates of these African people, as though they were a stack of Pokémon cards? One reason is that European explorers and historians had already begun to develop the idea of Africans as inferior and less than human. The first kidnapped Africans taken by the Portuguese were described by a royal chronicler as "bestial" and "barbaric." Hernando del Pulgar, Spanish historian, called Africans on the Mina Coast "savage people."[1]

Race, the problem of the twentieth century as so eloquently and timelessly stated by Dr. Du Bois, is in fact a competition, as discussed in chapter 5. Very few who are or were victims and losers of this race understood and probably still do not understand its true intent. Africulture's definition of racism mirrors Dr. Claud Anderson's definition of racism: a wealth and power-based competitive relationship between Blacks and non-Blacks. In his book *PowerNomics*, he argues that "true racism exists only when one group holds a disproportionate share of wealth and power over another group, then uses those resources to marginalize, exploit, and subordinate the weaker group."[2] The later, more familiar definition of race—a class or kind of people unified by shared interests, habits, or characteristics—determines what team you are on.

In the Americas, racism and anti-African prejudice predate the English colonies. Spanish colonizers introduced the concept of race in the Americas in order to justify their brutality toward Indigenous citizens of Turtle Island, some of whom were also descendants of early African explorers and had almost every shade of melanin in their populations.[3] The Taino, Caribs, and Arawak were some of the earliest victims of this new virus that ransacked and annihilated the peace and people of the fifteenth century Western hemisphere. Sawubona Tainos, Caribs, and Arawaks. (The people described as Caribs suffered misidentification as well as brutal savagery, as the label *carib* was a corruption of the Arabic term for amber.) In 1508, a Spanish priest named Bartolomé de Las Casas arrived in Hispaniola (the island that today is shared by the nations of Haiti and the Dominican Republic) and described the carnage that took place in those early years of pestilence:

> There were 60,000 people living on this island, including the Indians; so that from 1494 to 1508, over three million people had perished from war, slavery, and the mines. Who in future generations will believe this? I myself writing it as a knowledgeable eyewitness can hardly believe it.[4]

The blight spread as the first kidnapped Africans arrived in the Western Hemisphere on slave ships in 1501. Ponce de León set foot on Floridian soil in 1513, Hernán Cortés decimated the Aztec Empire and population in 1521, and Francisco Pizarro beheaded the thirteenth Incan emperor, Atahualpa, in 1533. The Portuguese and Spanish languages also spread the scourge of race through the lands they inhabited, explored, and annihilated when they began branding the Moors as Negroes. Negro then came to refer to not only a Black person, but a servant or slave. [5]

In what is now Georgetown County, South Carolina, the first European colony on the land that is now the United States, called San Miguel de Gualdape, was established in 1526 by a Spanish colonist named Lucas Vázquez de Ayllón. Ayllón ordered enslaved Africans in the colony to build housing for the Spaniards in June of 1526, which became the first recorded act of enslavement in the continental United States, sixty years before the landing of Europeans at Roanoke Island, eighty-one years before the Jamestown settlement, and ninety-three years before the *White Lion* landed at Point Comfort in Virginia on August 20, 1619, the first recorded ship of kidnapped, indentured, and soon-to-be-enslaved Africans brought to the newly established English colony.[6]

During this period was also the first documented case of what would later be known as drapetomania, a term coined by Samuel Cartwright in 1851 to describe a peculiar mental illness that caused enslaved African Americans to run away or try to escape for their freedom. These deranged freedom-seeking Africans revolted against the sickly, dying Spanish population, setting fire to the site of the colony, and fleeing into the forests to join the Indigenous communities that surrounded them.[7]

This blight spread into the next European colonial project at Jamestown, which didn't die out or phase out but rather metastasized into a soil and social cancer. Twenty-one years after the first enslaved Africans arrived in the English colonies in 1619, a Black man named Mr. John Punch, along with two white indentured servants, ran away from the property of Hugh Gwyn, a member of the Virginia House of Burgesses. All three were captured and were punished by thirty lashes with a whip. The two white servants had four years added to their period of indenture. But John Punch was punished by enslavement "for the time of his natural life here or elsewhere."[8]

This first documented case of drapetomania under British colonial occupation was at the beginning of a series of laws and acts in the British colony of America, which built a separate legal and social structure for Blacks and whites—the existence of slavery, of course, but also different punishments for crimes and all manner of restrictions on what Black people were allowed to do. This undercurrent meant that if you helped or assisted Black people to acquire their freedom or break the laws, acts, and statutes of the state, you would be punished in a manner differently. Even white colonial subjects or indentured servants would receive a harsher punishment for breaking laws if they were caught collaborating with Black people rather than acting alone. It was one of the many examples of the double standards related to the law and social order that people of Africa have had to navigate, and that would define the colonial era and the foundation of the United States of America. This double standard made unification between poor, indentured whites and poor, indentured, enslaved, or free Blacks less appealing. The goal of poor whites became moving into a higher class rather than freedom and justice for all. Dr. Du Bois would describe this phenomenon as the psychological wage of whiteness.[9] The payment of this wage is still one that many whites, poor or otherwise, seek, desire, and pander to today.

The main author of the Declaration of Independence, Thomas Jefferson, was himself a slave owner who had written that he suspected Blacks were "inferior to the whites in the endowments both of body and mind."[10] This thought process was inherently baked into the Constitution and thusly the foundation of the United States of America. The founders' deliberations had the oxymoronic intent of saying that all men are created equal, while considering only one demographic of person, a white landowning Protestant male, as *really* equal. This was cognitive dissonance on a national level.

In 1857, the opinion of Supreme Court Chief Justice Roger Taney in the Dred Scott decision summed up the unwritten law of the land. Not only were there 3 million people legally enslaved in America at that time, they had also, wrote Taney, historically been regarded "as beings of an inferior order, and altogether unfit to associate with the white race, either in social or political relations; and so far inferior, that they had no rights which the white man was bound to respect."[11] Blacks could never be citizens, had no right to sue, and had not been included in the famous phrase from the Declaration of Independence, "all men are created equal."

In Taney's day, the total value of enslaved people in America was 16 to 20 percent of all US wealth—the equivalent, in today's dollars, of $32 trillion. Those in power at the time were glad to have a reason to keep all that wealth where they thought it belonged. If Black people were inherently inferior and had no legal rights, that justified keeping them in bondage.

The Civil War was a brief moment of reckoning when the victims of US pestilence forced the nation's hand to confront its original sins. The Constitution and government had made the evil of slavery the law of the land, and US citizens had to fight to decide if all people should be treated humanely and not be subjugated to repression and tyranny by others. It was a war against the pestilence that was plaguing the soil, soul, and lives of the people of African descent in this nation.

The Civil War was a turning point in American history, as well as American agricultural history, shifting roles, paradigms, persistence, and policies that have sustained and stained American agriculture since 1861. It wasn't as simple

A CONSEQUENTIAL AMENDMENT

Just before the Civil War began, a proposed constitutional amendment known as the Corwin Amendment made its way through Congress, with support from President James Buchanan, stating:

> No amendment shall be made to the Constitution which will authorize or give to Congress the power to abolish or interfere, within any State, with the domestic institutions thereof, including that of persons held to labor or service by the laws of said State.

Although this amendment did not use the word *slavery*, its effect would have been to prevent any subsequent federal law banning slavery. Lincoln did not stand in the way of this amendment; in fact, in his first inaugural address, on March 4, 1861, he said, "I have no objection to its being made express and irrevocable."

An insufficient number of states ratified the amendment, though, and it was never adopted, even after the Confederacy adopted their constitution in Montgomery, Alabama, on March 11, 1861. Adoption of the Confederate constitution was a major line in the sand, as the South was the breadbasket of the United States and could potentially starve the North of vital food, raw materials, and textiles. Lincoln's support of a constitutional amendment that would have prevented the abolition of slavery is a contrast with his image as the Great Emancipator. As he stated on many occasions, he was more concerned about keeping the Union together then he was about ending the immoral institution of slavery.

as the good abolitionists and Northerners, led by a saintly Lincoln, stopping the backward Southerners, however. And Black people played a much more significant part than just watching passively from the sidelines—there was Black resistance at work long before the Emancipation Proclamation, emanating from farms and plantations from the north to the south.

Exodusters. *Photo by Missouri History Museum / Wikimedia Commons.*

For the Black farmers trying to survive after the end of slavery, the Civil War era was a mixed bag. Alongside emancipation, that time brought in a new group of structural disadvantages to being Black.

The famous 1862 Homestead Act offered citizens (as long as they had never fought against the Union) nearly free land, as long as they lived on it and raised crops. Over time, 270 million acres would be given away under the Homestead Act.[12] The first problem with this program, for Black people, was that they weren't citizens—at least not until the Fourteenth Amendment was passed in 1868. But the bigger, long-term problem was that you needed capital and/or community cooperation and some level of government approval in order to improve a homestead. If you were destitute, you weren't going to be able to afford to move your family, build a house, and buy seed and a plow and fences and livestock. All that cost was, on average, about $1,000 at the time—$17,500 in 2020 dollars.[13] If you didn't have the money, you needed the hands of your family or the community to make it through the arduous winters and survive agriculturally. Feeding your family was top priority. Many, like the Black Exodusters, had to live in dugouts—literally holes in the ground—in places like Nicodemus,

Kansas, where the Indigenous nations such as the Osage assisted settlers in their former homelands.

It wasn't just hard to get started; it was hard to keep covering your costs as time went on. Homesteading happened most often on the Great Plains, in arid places that were notoriously difficult to farm. Dr. George Washington Carver himself tried homesteading in Kansas and ended up forfeiting his claim because of trouble with his crops. Of about 1.6 million claimants who ended up being successful with their claims, only about 3,500 were Black.[14] That's approximately two thousandths (0.002) of 1 percent of the total beneficiaries. The challenge in many cases, very much like today, wasn't always growing crops, but finding a market for them once they were grown. Having a purchase-oriented community is one of the keys for a successful agricultural enterprise, either as a homesteader or a farm.

The Homestead Act was America's largest welfare program, and when evaluated in terms of the value of the land given away, it was one of the greatest wealth transfers in the history of the world, benefiting 93 million homesteader descendants alive today, using land taken from Indigenous nations of the United States and Mexico. The land value today is equivalent to providing a family a $500,000 asset that was increasing in value every year, and could be used to leverage capital. To claim it, all you had to do was be considered white and live on the property for five years. The land that was stolen and given to the first government welfare or land recipients would be valued today at between $900 billion and $2 trillion dollars.

In 1864, concessions to the African American population were made when General William Sherman and US Secretary of War Edwin Stanton met in Savannah, Georgia, with approximately twenty formerly enslaved religious leaders to discuss the best way forward. These clergy and community influencers communicated their interests in land and the ability to farm and support themselves. Special Field Orders No. 15, the legislation that grew out of these meetings, would have carved up 400,000 acres of Confederate plantations to resettle the emancipated, affording every Black family the now infamous "forty acres and a mule." The land was going to spread from Charleston, South Carolina, to St. Johns River, Florida, including the sea islands of Georgia, and the victorious military had a surplus of mules that would have transitioned to civilian life as service animals. Approved by President Lincoln, this act was one of the many government-sanctioned disappointments that Black families and Black farmers would face as they tried to place their faith in the American government for the first time.

It was—and may still be—naive of people of African descent to trust the United States government to have their best interests at heart. After Lincoln was assassinated and Andrew Johnson became president, Sherman's order was quickly rolled back, along with the power of the Freedmen's Bureau, a federal agency founded in 1865 to assist freed Black people, to hand out acreage. Representing the state of Tennessee, Johnson was a sympathizer to the Southern cause, and the standards of the Southern social system. His actions indicated that he was actively opposed to people of African descent being treated and valued with the rights and liberties of any American citizen. Johnson wanted to protect private property and was not sympathetic to empowering Black people. It is suspected that Johnson was a sympathizer of the secret organization the Knights of the Golden Circle, whose aim was to create an agricultural empire fueled by slavery. In 1868, he vetoed Senate Bill 60, which would have made the Freedmen's Bureau a permanent federal institution. The Freedmen's Bureau was bankrupted by a corrupt board of directors, and by 1872 it was entirely defunct.

If every emancipated person had actually received forty acres, by the way, that would have represented an $800 billion value in today's dollars.

Another short-lived program, the Southern Homestead Act of 1866, was intended to help newly emancipated Black people (and Southern whites who had been Union sympathizers) become landowners in the South. But it was a failure. The 46 million acres designated for these homesteads was mostly forested or swampy—not suitable for growing row crops. Black people found themselves ineligible if they were already under wage-labor contracts negotiated by the Freedmen's Bureau. They were pushed further to the side when the act was expanded in 1867 to include former Confederates and any Southern-empathizing white citizens.

Many newly freed people had nothing to their names, and that poverty prevented them from establishing farms. Not to mention the violent racism of the Reconstruction South. Some reports state that approximately 6,500 free Black citizens applied for land claims but only 1,000 managed to prove claims under this program.[15] Notably, though, Black homesteaders were more likely to be successful as homesteaders than whites—after all, they were the farming experts in the South and historically were able to build from the ground up with limited government or social support, oftentimes in hostile environments.

Only a decade after it was passed, the Southern Homestead Act was repealed to open that land to Northern timber companies, setting the stage for the sharecropping economy. The social climate wasn't designed to allow for the disenfranchised, 100 percent–taxed class of people who built this nation to own much land, no matter its condition or how it was acquired.

Land ownership is key to American farming, but so is governmental support. It was also Lincoln who established the United States Department of Agriculture with the passing of the Organic Act of 1862, during the middle of the Civil War. This was the first time the government had allotted federal resources to shore up the most critical industry of the nation. Such action would affect the nation's future greatly, no matter who won the war. Lincoln shared with the nation, in his December 3, 1861, State of the Union address, that he believed agriculture to be "the largest interest in the nation"—essential to the economy—and he thought it should have a proper governmental department to oversee it

As Honest Abe's address implied, the agriculture industry had been somewhat independent prior to 1861. I'm sure you can guess why. The potential loss of its priceless human capital imperiled American agriculture and was the impetus for creating the Department of Agriculture in the United States.

The USDA would support farmers, who in turn were feeding the Union Army. But vestiges of the plantation system remained in the USDA after the conclusion of the Civil War. For the first few decades of its existence, its programs failed to include one specific class of farmers. (I'm sure that after reading this far, you don't need me to tell you who that class was.) The agency saw (and sees) the existing labor force, no matter the era, as expendable and exploitable, whether it is sharecroppers, white small farmers, migrant workers of various national origins, or, in the twenty-first century, temporary and seasonal H-2A workers.

BLACKLASH

There may have been some hope that America would attempt to right the wrongs of its original sin. Because of General Lee's ultimate surrender at Appomattox Court House in Virginia and the valiant fighting of the United States Colored Troops, there were some ideas about how America would now treat its citizens of African descent as citizens after the Civil War. However, an expectation of Blacklash was ever present. The public relations campaign of the United Daughters of the Confederacy and other sympathizers of the Confederate cause named the war as the War of Northern Aggression, and attempted to convince the public that they were fighting to preserve states' rights, rather than to preserve the practice of enslaving human beings for their personal profit, benefit, and pleasure. They conveniently left out that by firing on and attacking Fort Sumter, a US military installation in South Carolina, on April 12, 1861, the confederate cause forced a civil war, turning heated debates and quarrels in Congress into bloodshed, gun fire, devastation and death across the Southern American landscape. This pattern of backsliding

During the Civil War, Black people served as soldiers in the US Colored Troops, as sailors, and as drummers, among numerous other essential roles. *Clockwise from top left: Photos from the United States Sanitary Commission Records, New York Public Library / Wikimedia Commons; National Archives / Wikimedia Commons; Ball & Thomas / Library of Congress.*

continues to repeat in American history, even in 2025 as I write this book, 160 years after the culmination of the War of Southern Aggression.

Blacklashing started with the treatment of Black Civil War veterans. A total of 209,165 African Americans were enlisted to fight in the Union Army, with thousands of additional soldiers and espionage experts serving for their freedom as well. More than a quarter of the sailors in the Union Navy, approximately 23,000 sailors, were Black—both male and female sailors. Less than 1 percent of the North was Black, yet African Americans made up over 10 percent of the army and about 23 percent of the navy. It was these valiant and courageous soldiers and sailors who turned the tide of the war and allowed for a Union victory. The United States Colored Troops earned twenty-six Medals of Honor. One commanding officer was so impressed with the troops' courage, tenacity, and outright dog in them that he decorated them at his own expense. Major General Benjamin Butler had 197 medals made by Tiffany & Co. of New York to award to his USCT troops after the Battles of Chaffin's Farm and New Market Heights, because the military itself did not honor them.[16]

There were numerous acts of heroism, which history has overlooked, by Blacks who served as spies, saboteurs, fighters, medical support, and information specialists who never officially enlisted in the Union Army but played significant roles. Black soldiers fought in more than 200 battles and skirmishes, including the Battle of Fort Wagner, the Battle of Chaffin's Farm, the capture of Richmond and the Battle of Appomattox Court House. Even children served, including as drummer boys—their average age was fifteen, and Elijah Mason served as a Union drummer at the ripe old age of eight.

As mentioned in chapter 5, agriculture played a pivotal role in the Civil War. Throughout, the fight was for land control, and especially the last two years of the conflict were focused on provisions and African people, as well as African crops such as coffee, cowpeas, and peanuts. General Sherman, who some historians have noted as racist based on letters he wrote to his wife, was victorious in his 1864 Atlanta campaign, thanks to the protection of his supply lines, which consisted of food and rations, shoes, artillery, and medical supplies. He attributed his victory to this protection, and personally placed the 14th United States Colored Troops to guard his supply lines in Dalton, Georgia, in August 1864.[17] Ironically or prophetically, troops guarded supplies that came from Africa, where their Liberian countrymen were aiding and supporting their fight for their literal freedom. During the war, if I'd been around to provide a coffee commercial, I'd have touted the best part of waking up and fighting for your freedom: Liberian coffee in your cup.

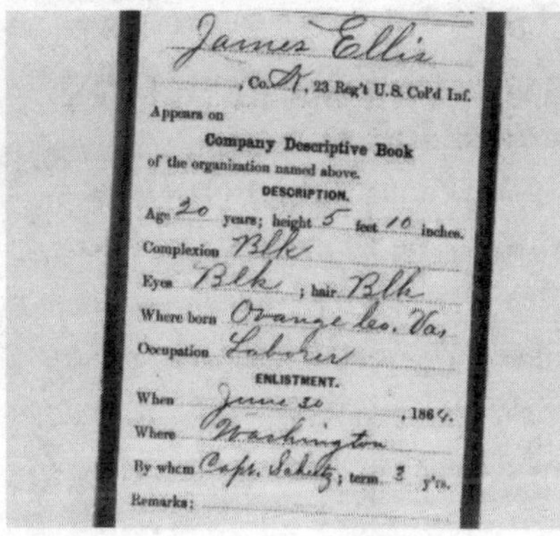

James Ellis
, Co. K, 23 Reg't U.S. Col'd Inf.
Appears on
Company Descriptive Book
of the organization named above.
DESCRIPTION.
Age 20 years; height 5 feet 10 inches.
Complexion Blk
Eyes Blk ; hair Blk
Where born Orange Co. Va.
Occupation Laborer
ENLISTMENT.
When June 30, 1864.
Where Washington
By whom Capt. Schultz; term 3 y'rs.
Remarks:

My fourth great-grandfather Mr. James Ellis served in the US Colored Troops.

Quite a few of the enslaved fought with and sometimes for the Confederate side, some against their will, others freely. Some were loyal to the conditions of enslavement and believed the propaganda related to Blacks as not being able to govern themselves and to how good and kind their "masters" were. Others took the opportunity to sabotage or provide intelligence for the Union side, not caring what side won as long as they would have an opportunity to be free as a citizen in the country. There were another approximately 6,000 to 10,000 Black Virginians who were body servants on the Confederate side, giving their dignity and/or lives as logistical support—many managed to provide counterintelligence while they were at it.

I have a unique, or maybe not so unique, connection to the Civil War, as I have numerous grandfathers and relatives who fought in the War of Southern Aggression on both sides. My fifth great-grandfather, William A. Jones, my grandma Lizzie's great-grandfather, served for about sixty days in the 13th Infantry Regiment of the Confederate Army.

Many Black Union soldiers and sailors and Black Confederate fighters and support staff would face discrimination when they tried to claim pensions and disability payments after the war. Wounded white veterans received benefits at twice the rates of wounded Black veterans.[18] Black soldiers often lacked the documents needed to apply—they didn't have birth certificates because they'd often been born into slavery, and they often didn't have hospital records because Black war hospitals were understaffed and couldn't accept many patients. Not only that, complaints from white veterans of invisible problems like back injuries were believed more often compared with Blacks.

Mrs. Harriet Tubman, the first woman to ever lead a raid, or three, in US military history, spent nearly three decades fighting to receive a veteran's pension for her Civil War service. She was ultimately given a widow's pension after

AN AFRICAN-INSPIRED NEW YEAR'S TRADITION

No matter where I'm at in the world, on January 1, I eat black-eyed peas, and every year my mother has called me to make sure I have some. The love of a mother, a grandmother, a great-grandmother, and a great culture has been passed down one bean at a time for at least the last 162 years.

The tradition of eating black-eyed peas on January 1 started during the late hours of December 31, 1862, after the first watch night service was held by those enslaved and freed Africans who anxiously waited to fight legally for their own freedom on January 1, 1863, the date the Emancipation Proclamation went into effect. In my family, the dish has been prepared through the generations by my mother, Mrs. Ann Carter, Grandma Lucille, and Grandma Rose as well as my great-grandmothers, Gran, Grandma Lizzie, and Nana as a mandatory tradition for good luck and prosperity for the year. Their black-eyed peas dish could not be eaten without mixing in some stewed tomatoes or tomato pudding, a mixture of bread (sometimes stale), tomatoes, and sugar into a confectionery delight, only served on January 1. Until I started cooking myself, I never ate this any other time of the year. The stewed tomatoes have always been my favorite part, as they pair well with the beans, corn bread or spoon bread, and some collard greens or other seasonal brassica greens like mustard or turnip.

While in Ghana, I had numerous favorite dishes, including egusi with nkotomere, groundnut stew with corn bread, and of course red red. When cooked well, red red had me try to bite through my spoon a few times because it was so good. This dish consists of black-eyed peas (or cowpeas, as they are called in Ghana), cooked in palm oil, and served with boiled or fried "red," meaning very ripe, plantain. The plantain in this stage is at its peak of ripeness, naturally releasing ethylene, a gas that changes its skin color from green to yellow to black, and makes the fruit considerably softer.

Red red derives its name from the deep burgundy color of the palm or zomi oil (a type of palm oil specifically from the Ewe people that's salted or spiced and has a different flavor than plain palm oil), the tomato paste and tomatoes, as well as the plantains, which are "red" in name only. Like the black-eyed peas that the African American community and now the larger community traditionally eats on New Year's Day, red red is culturally associated with good luck and prosperity.

I've postulated, through spirit and connection, that red red potentially was a dish carried in the ancestral memory of some of the West African captives. No official record exists of when we started eating black-eyed peas and tomatoes, but both items were considered slave food at that time. It could be prepared and consumed without much fuss, and black-eyed peas kept well through the winter. When the opportunity came for the newly emancipated masses to fight for the freedom of all, they did just that, but before they set out, those motherly elders prepared them a dish for luck, prosperity, and success. This was their version of red red—stewed black-eyed peas with the red tomatoes. The enslaved didn't have access to palm oil or plantains, but their version stuck as a tradition, and collard greens were added later. Where tomatoes weren't available, they were left out, and Hoppin' John would be prepared instead.

her veteran husband's death instead of her own and it took a decade for her and her supporters to get her widow's pension raised from $8 a month to $20 a month in 1899, proving that discrimination didn't always discriminate even for notable and historic heroines like Mrs. Tubman.[19] My fifth grandmother applied for a widow's pension because my fifth grandfather, Mr. William McIntosh, the great-great grandfather of Gran and Grandpa Gil, fought in the Union Army, in Company B of the 27th Missouri Infantry. He was between the youthful age of seventy-six and eighty at the time. Because he fought for the Union Army and not the US Colored Troops, he was deemed white by society, and his wife, Polly McIntosh, who I believe was a white woman slave owner,

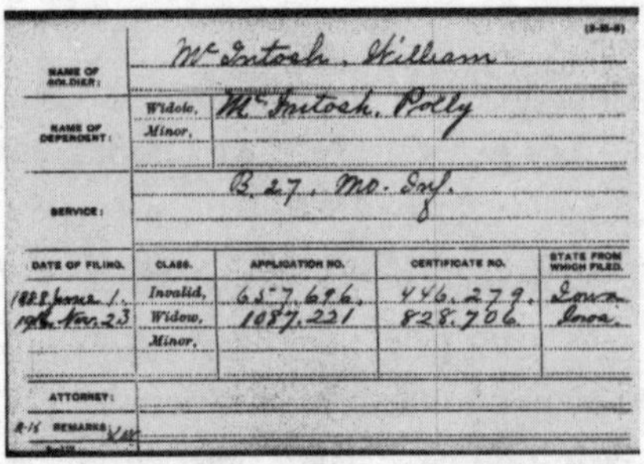
NAME OF SOLDIER: McIntosh, William

NAME OF DEPENDENT: Widow, McIntosh, Polly
Minor,

SERVICE: B. 27. Mo. Inf.

DATE OF FILING.	CLASS.	APPLICATION NO.	CERTIFICATE NO.	STATE FROM WHICH FILED.
1888 June 1.	Invalid,	657.696.	446.279.	Iowa
19[illegible] Nov. 23	Widow,	1087.221	828.706	Iowa
	Minor,			

ATTORNEY:

REMARKS:

Mr. William McIntosh's pension record.

was awarded a widow's pension in 1888, a year before Auntie Harriet. Sawubona Mrs. Harriet Tubman.

More resistance came from the free and enslaved leading up to, and during, the Civil War than is often taught. Many resistance fighters of enslavement are not widely known, and that includes the activities of the clandestine organization referred to as the Lincoln League (also called the Loyal League or Legal League)—a secret organization that pushed for the freedom of enslaved people. The league began as early as the 1790s, when Prince Hall, a free preacher in Pennsylvania, started a division of the Freemasons devoted strictly to African Americans.

The Loyal League was born out of a longstanding thread of uprisings and rebellions: the Stono Rebellion in South Carolina in 1739, the Haitian revolution in 1791, Gabriel's Conspiracy in Virginia in 1800, the revolt of 1811 in Louisiana, Nat Turner's rebellion in 1831, and offers of freedom from the British to Black soldiers who crossed the line during the War of 1812.*

The members of the Loyal League risked their lives to try to free their enslaved mothers, fathers, sons, daughters, brothers, and sisters. The Loyal League's leaders were a mix of free and formerly enslaved African American men and women. These lesser-known founding fathers and mothers stated that they would not be a part of a rebellion. Rather, they desired another *revolution*, a nationwide insurrection in league, in agreement, in line with the Constitution of the United States and its principles and overarching themes. The leaders included Dr. John S. Rock, a Harvard-trained lawyer and one of first Black lawyers in the United States; Mr. William Howard Day, an educator and minister; Mr. Frederick Douglass (who received crucial support from his wife, Mrs. Anna Murray Douglass, an abolitionist and Underground

---

* This last bit of history is why the third verse of "The Star-Spangled Banner," which nobody ever sings, describes how "no refuge could save the hireling and slave"—from Francis Scott Key's perspective, the enslaved people were the turncoats, and their freedom was not penned in that song.

Railroad organizer); and Major Dr. Martin Delany, who founded a newspaper called the *Pittsburgh Mystery* in 1843. Sawubona.

The Loyal League held conferences, such as the National Emigration Convention of Colored People, led by Dr. Delany in Cleveland in 1854. The Black leaders discussed, planned, and strategized: "How do we get our people out of bondage?" At a subsequent conference in 1858 in Chatham, Ontario, John Brown made a case for his form of rebellion, encouraging the type of violence and uprising Mr. Brown had participated in Kansas at the Pottawatomie Massacre in 1856. But his plan of having Africans take up arms against the US government was not seen as practical or in league with the US Constitution.

In response to this rejection, John Brown called Dr. Delany a coward during the Chatham conference.[20] Brown couldn't understand that the Loyal League had been making plans and arrangements for years to support their goal, such as working with people on plantations who were serving as intelligence officers. These operatives were taking remarkable risks to provide information from the highest levels.

The leaders of the Loyal League were excellent "chess players" who had a laser focus on pushing forward their agenda, and they used Brown as a pawn

Dr. John S. Rock. *Internet Archive Book Images / Wikimedia Commons.*

Major Dr. Martin Delany. *Photo by Peteforsyth / Wikimedia Commons.*

Mrs. Anna Murray Douglass. *Photo by Rosetta Douglass Sprague / Wikimedia Commons.*

to initiate the war. They never intended to participate in his scheme to raid a military arms depot in Harpers Ferry, Virgina, but they saw that his actions would create so much paranoia and discontent from the Southern perspective that it made the idea of a free Black population impossible. Meanwhile, from a Northern perspective, the John Brown affair planted a seed in the minds of the Yankees that white men could and should fight on the behalf of this country's enslaved population and the abolition of an immoral and unethical practice. While Brown's raid at Harpers Ferry was going on, all of the Loyal League leaders were out of the country. They didn't want to be associated with this uprising at all.

In 1866, Congress passed the Civil Rights Act, which was supposed to protect all people born in the United States by granting them citizenship, "without distinction of race or color, or previous condition of slavery, or involuntary servitude." President Andrew Johnson vetoed the act, which was then passed by a congressional override of the veto.[21] Congress was feeling strongly about this at the time—it was the first congressional override of major legislation in history. Parts of the act were echoed by the Fourteenth Amendment a few years later.

Yet the enforcement of this law was severely lacking. It was like trying to treat a crop's leaves or fruit to stop a disease when the problem arises from poor soil full of aggressive, negative elements. The Republican lawmakers who had pushed through this original Civil Rights Act found it wasn't politically advantageous to stand behind it and enforce it, so they backed off. And state-level Black Codes effectively trumped the federal law, making racial matters a states' rights issue. It took almost a full century for the *other* Civil Rights Act, the more famous one, to come along.

The 1870s was a decade of disappointments for the newly emancipated. Political and economic pestilence were widespread. The Freedman's Savings Bank was created to allow Black people to put their money in a United States bank for the first time, but it was run exploitatively by Jay and Henry Cooke, brothers who were investors and financiers of the Union during the Civil War. The bank should have just been called the Cooke Brothers Failed Railroad Investment No Accountability Slush Fund, but I'm sure Freedman's Savings Bank was a bit more catchy.

The Freedman's Savings Bank collapsed due to fraud and mismanagement, resulting in the loss of around $3 million to Black depositors while enriching white managers. Frederick Douglass, who tried to save the bank at the eleventh hour, wrote that it had become "the black man's cow, but the white man's milk."[22] No one went to jail; there was no hearing, no trial, just millions of dollars stolen from people who entrusted what they had, in faith and in dollars, to the US government and its institutions, and were snubbed. Dr. W. E. B. Du Bois was appalled at the setback to those who had been ruined and the Black community in general: "Not even ten additional years of slavery could have done so much to throttle the thrift of the freedmen as the mismanagement and bankruptcy."[23] The disaster made many African Americans highly distrustful of banks and financial institutions. Keeping money under mattresses or floorboards or in cans, stockings, bras, and socks became the safest choice.

Federal troops were pulled from the South as part of the Compromise of 1877. These consisted of large delegations of Black troops whose responsibility had been shifted to federal law enforcement related to discrimination and violence against America's new Black citizens in the former states of the Confederacy. The Compromise of 1877, which ended Reconstruction, allowed Rutherford B. Hayes to assume the presidency after a contentious election. Withdrawing the last of the federal troops from South Carolina, Florida, and Louisiana shifted power and prejudice back to the white citizens of the South, leaving no federal protections for Blacks against their terrorism, intimidation, and, later, state-sanctioned (or ignored) violence. It allowed for what the Confederate states stated that they desired: states' rights, which meant imposing state-sponsored discrimination and subjugation if slavery was no longer an option. A meeting at the Wormley Hotel in Washington, DC, on February 26, 1877, between representatives from Hayes's Republican camp and representatives from Democratic candidate Samuel Tilden's camp settled the disputed 1876 election between Hayes and Tilden. It also set back the African in America fifteen years when the Reconstruction era ended, allowing states to legally discriminate and break the law of the Civil Rights Act of 1866. All of this paved the way for the American apartheid era. It's ironic that the decision to usher in a new gilded age of segregation, and separate but unequal policies and standards for the country, would be negotiated in the lobby of a hotel owned by Mr. James Wormley. A quite successful African American entrepreneur, Mr. Wormley's exemplary service and farm-to-table cuisine made it one of the most lauded hotels of the time in Washington, DC, and it hosted many of the most influential and powerful people in the country.

The Compromise of 1877 marks the beginning of what is often called the Jim Crow era. But that name doesn't truly capture the nature and spirit of the era. Jim Crow was a fictitious minstrel show character that Thomas Rice, among other white minstrel performers, made popular as a depiction of America's Black population. "Apartheid" has a different type of permanency. The original definition of the word in Afrikaans simply means "separate." And that's what America in this era embodied. Just like in South Africa, this apartheid was a separate system that gave rights and benefits to one group and denied them to the other, very publicly, making this the socially, religiously, and scientifically justified standard of

Thomas Rice performing in blackface. *Photo by University of Illinois Urbana-Champaign / Wikimedia Commons.*

society. "Jim Crow" doesn't do justice to the realities of this era. American apartheid or the blackface era are less sanitizing terms and should be used to describe what happened for the next seventy-five years as America both industrialized and further racialized its society, movements that developed hand in hand. Apartheid was another pestilence. Society was the soil, lynching represented locusts and other leaf-destroying pests, psychology was a harsh wind that bent and broke the young plants, and policy was a caustic acid rain that led to degeneration while ostensibly encouraging growth.

Apartheid in America meant the banning of mixed marriages, restrictions on Black people's movement, and the denial of their access to public spaces and government programs. There were restrictions on Black people's ability to bear arms, buy property, use banks and lawyers, and buy or sell agricultural products, especially to white people. Many businesses wouldn't sell to Black consumers, and many more wouldn't buy from Black producers. White landowners, who had always depended on Black labor, did everything they could to keep workers under their thumb. Slavery had been outlawed, but the system kept the emancipated from living like free people. They were prevented from voting and became trapped in coercive labor arrangements. White supremacy and Justice Taney's haunting ruling in *Dred Scott v. Sandford* were the de facto law of the land.

Vagrancy laws were one of the most pernicious elements of the Black Codes. It was a crime, if you were a Black man, to be out of work, or doing a job that whites didn't recognize as work. If you failed to pay certain taxes or carry a work contract, you could be accused of vagrancy. And since the Thirteenth Amendment, ratified in 1865, had outlawed slavery except as punishment for a crime, the convict leasing system arose as a different way to harness the work of Black men imprisoned for vagrancy and other offenses, as defined by the white power establishment. Under these pressures, many Black families in the South were "caste" into the sharecropping system on land owned by the people who had previously owned them—or on nearby plantations. Independence gave way to dependence on working on a farm to try to make ends meet.

Violence was the lifeblood of this system, perpetrated by groups like the Ku Klux Klan and by vigilante "justice" carried out without fear of legal prosecution or punishment. Between 1875 and 1955, more than 4,700 people were lynched in the United States—that's equivalent to one lynching per week for eighty years.[24] This is was what I call the Black Scare, like the Red Scare of the 1950s. Black people were made out to be insatiable beasts, with little or no conscious self-control, who were rabid predators that preyed on white society. Popular culture cemented this perception through one of America's earliest hit films, *The Birth of a Nation*, which was screened at the White House by a Virginia-born president, Woodrow Wilson. This film pushed the premise that Black men were raping white women. Wilson loved the film and raved about it, put his presidential stamp of approval on it, and called it a very accurate depiction of Negroes. This propaganda only supported and enforced Wilson's policies to segregate the federal government and limit the employment of Black workers in the government to menial roles.

Because of that propaganda and fear, innocent men, women, and children were lynched. Babies were cut out of pregnant mothers' stomachs, some hanged with their own umbilical cords, others stomped to death, as happened to Mrs. Mary Turner in Lowndes County, Georgia. One day after she adamantly and rightfully spoke out against the lynching of her beloved husband, Mr. Hazel "Hayes" Turner, on May 18, 1918, she was hanged upside down from a tree, doused with gasoline, and her clothes burnt off; her baby was cut out of her stomach and stomped to death by her brutalizers, and finally she was shot over a hundred times. Her death occurred because her lynched husband, Mr. Hazel Turner, a sharecropper, was accused of participating in the murder of Hampton Smith, a white farmer. There were seven confirmed lynchings associated with the death of Hampton Smith; they were directly tied to agriculture, sharecropping, vagrancy laws, and the audacity to be Black and stand for the rights of yourself and the ones you loved. These were horrific crimes, often perpetrated in public, and totally unpunished.[25] Sawubona Mr. Hazel Turner, Mrs. Mary Turner, and unborn baby Turner.

Lynching put every Black person on notice that their lives were in danger from random, vicious violence. A terrifying pestilence, for sure. Very few, if any, people were tried for these crimes, and no one was found guilty of lynching a Black person—a public kidnapping, assault, and execution without due process by a jury of their peers. "Justice" Roger Taney's 1857 words were being acted out as law: that a Black man or woman had no rights a white man was bound to respect, not even the right to live in peace. There was an intentional failure and foregoing of public and civic duty to enforce

Dr. Ida B. Wells-Barnett. *Photo by Yann / Wikimedia Commons.*

laws at any local, state, federal, religious, moral, or human level to protect the lives and rights of African Americans, who were viewed at this time more like vermin that needed to be exterminated than citizens who needed to be respected, despite their continued contributions to agricultural and American society.

As far back as 1892, Dr. Ida B. Wells-Barnett was campaigning for anti-lynching legislation. Sawubona Dr. Ida B. Wells-Barnett. Guess when America finally got a federal anti-lynching law? 2022. I could have written, "Guess when we," but "we" means that you feel like a part of something. And for most of my life I have not felt a part of this nation or a citizen of this nation, because of what society has tried to do to the kidnapped and culturally amalgamated population that I am descended from.

Out of touch media pundits and their audiences cluelessly wondered why there was such an uproar in May 2020 after white authority figures supposedly carrying out the law murdered George Floyd. I believe Floyd's killing triggered our collective genetic memory of our people getting killed. It evokes the injustices and pain suffered by Mr. and Mrs. Turner and their unborn child, by Mr. Frank Embree, Mr. Will James, Mr. Joseph Richardson, Mr. Emmett Till, and the 200 Black people beheaded in Southampton County, Virginia, in 1831. Blackhead Signpost Road is named after this horror. There are so many victims, and the pain courses through the platelets, mitochondria, and cellular membranes of our beings. As the incomparable Mr. James Baldwin etched into my medulla oblongata, to be Black and conscious in America is to be in a constant state of rage. I've been there, in a constant state of repressed rage, since 1981.

Reminders of these atrocious experiences can arise during a breaking news segment, in content on social media, or while reading books about

Thousands of people were lynched in the late nineteenth and early twentieth centuries. *Photo by J. Horgan Jr. / Wikimedia Commons.*

The Elmina slave dungeon in Ghana. *Photo by Okletey Martins / Wikimedia Commons.*

black experiences in this country. But this ancestral memory is even older than the country we call America. What I and many saw in the beating of Mr. Rodney King, the brutalization of Mr. Abner Louima, and the killings of Mr. Amadou Diallo, Mr. James Bell, Mr. Mike Brown, Mr. Trayvon Martin, Ms. Breonna Taylor, and Mr. Ahmaud Arbery was a reminder of the terror experienced by African people in the first part of the passage to America. The journey of terror didn't start with the Middle Passage, but with assault or murder in your own home or community. If you were truly unfortunate, you were kidnapped—in the middle of the night or in broad daylight—and chained together with other captives to walk, sometimes upward of 250 miles, through bush and forest paths to a coastline you may have never seen or known existed. Inside a massive building, you may have laid eyes for the first time on people with pale skin who wore funny clothes and spoke a foreign tongue. You were thrown in a dungeon, where you'd hear the screaming, crying, and pleading from people trying to communicate their pain, suffering, anger, or disdain to ears that didn't even care to acknowledge your existence, know your name, or care about the freedom, liberty, and life being stripped from you and your progeny when it didn't directly serve and benefit them.

I've listened to tour guides leading groups through these dungeons many times—I've led tours there myself. These slave dungeons, or castles as some called them, had the most luxurious amenities—stone walls and one small opening that served as ventilation in a twenty-by-ten-foot space, holding upward of 250 strong, muscular, furious people. The ones who survived the trek were given two to three handfuls of water and one handful of food as their provisions each day. Two to nine weeks of living in these conditions created a literal survival of the fittest. I failed to mention this amenity: the bathrooms, which were the ground on which you stood. This led to a permanent layer of feces, urine, mucus, menstrual blood, and vomit eighteen to twenty-four inches deep, that you had to wade through and stand in daily, barefoot. The space was so tight with people that laying down was impossible unless you passed out, so you had to sleep and rest standing up or leaning against someone or the wall. And the smell—you can just imagine the smell. The only time the floors were "cleaned" was a few times a year, when it rained enough for water to run through the slave "castle."

We, the descendants and ascendants of these survivors, carried this in our seeds, and we planted these seeds of unparalleled horror in the American, Caribbean, and South American soils we were dispersed and lodged in. We utilized this ancestral memory to strengthen our souls for the enslavement, apartheid, and postapartheid eras of America. This terror and future trauma is the compounded interest of racism and discrimination. It persists because of the constant accrual of negative images and aspersions around people of African descent, a phenomenon that is part of a war of terror on melanated and Indigenous people the world over.

SEPARATE AND UNEQUAL

Despite the threat of random violence and the various governmental obstacles stacked against them, the number of Black farmers increased steadily in the decades that followed the Civil War. Black families managed to accumulate land equivalent to the area of South Carolina by 1910, and they owned 925,000 farms—in part because of the help of Black institutions.[26] But federal policy makers stepped in to legally exclude Black farmers. As noted in chapter 2, the 1914 Smith-Lever Act excluded Black farmers from extension services, and they were unable to benefit from the farm credit system and farmers co-ops established by the 1916 Federal Farm Loan Act.

As the Great Depression hit, the government went into overdrive with the New Deal, passing legislation and creating programs meant to jump-start the economy and help out select citizens to create the middle class. But again the help was unequally given. The Agricultural Adjustment Act of 1933 was supposed to stabilize crop prices, in many cases by reducing production. This was the nation's first subsidy, paying white landowners not to produce crops. While these landowners benefited, the policy hit Black tenant farmers and sharecroppers in the South like an earthquake, taking away their livelihoods. Their landlords blocked direct relief for the tenants, arguing that the Black workers wouldn't know how to handle the money—a trope, still heard today, that stems from a misinformed perception of the causes of the bankrupting of the Freedman's Savings Bank. Additionally, the landlords often used the subsidies to purchase equipment that made Black workers unnecessary, driving American agriculture toward larger, more mechanized farms and pushing many Black tenants and sharecroppers off of their farms.

Farmers and domestic workers, the majority of whom were Black, were excluded from the National Labor Relations Act of 1935, which protected unions. And farm workers weren't covered by the Fair Labor Standards Act of 1938, which established a minimum wage and outlawed child labor.[27] Even the Social Security Act of 1935 excluded farmworkers (and, again, domestic workers). "Agriculture was different," people argued. But farmworkers were left without the protections and benefits that most other American workers enjoyed.

As a Black farmer in this period, you faced the probable reality of needing to work until you died, with no social safety net for you or your family. By incorporating your farm as a business, you might be able to access some benefits. But in the 1930s, you had to hire a lawyer to help set up a business, and how many Black lawyers were there in the United States at this time? Mr. Thurgood Marshall could not take time to prepare a corporation filing for you, or write you a proper will, while also fighting for integration in American public schools and dealing with lynching cases across the length and breadth of the nation. And if you went to a white lawyer's office, you could be accused of a crime under the Black Codes.

DEMON CANNABIS

Certain species of plants were also demonized and criminalized in the 1930s. Prohibition of alcohol ended in 1933, but a new crop with African roots became the target of scandal. The Marihuana Tax Act of 1937 made illegal the possession, distribution, sale, or transfer of cannabis, with the exception of certain industries. In 1942, tax stamps were issued to farmers to grow hemp and aid in the war effort. Four hundred thousand acres were grown, cultivated, and sold to the US government.

The Bankhead-Jones Farm Tenant Act of 1937 was supposed to help tenant farmers by giving them long-term, low-interest government loans to buy land and equipment. But fewer than a thousand loans were made to Black farm tenants. Local county officials made the decisions about lending, and they weren't looking to help Black people. Most applications that Black farmers submitted were delayed, denied, or offered only on unfavorable terms.[28] The USDA made the local committees permanent in 1953, naming them Agricultural Stabilization and Conservation committees, made up of three to eleven members who serve three-year terms. The county committees were local white farmers who could, and oftentimes would, make recommendations for or against farmers and potential USDA program participation. The committee system placed a new, very personal and direct hurdle to Black farmers. The county committees served as the gatekeepers to government programs and support. The committees sometimes made vengeful, spiteful, and opportunistic decisions, which often led farmers to abandon efforts to try to work with the USDA out of sheer frustration with the county committees.

For thousands of Black farmers, the economic realities of the Great Depression were too much to bear without the government support that white farmers were being offered. This and several other factors all fed into the mass movement of Black people from the rural South during the Great Migration. As more people moved north, they often found themselves redlined as the Federal Housing Administration—another New Deal agency—refused to issue mortgages, failed to provide financial assistance, and cleverly discriminated on the basis of zip code.

The GI Bill, as discussed in chapter 2, did little to help Black veterans as compared to whites. By 1956, 8 million veterans had gotten educated under the GI Bill and $33 billion in home loans had been issued. But very little of this assistance had gone to Black people—and it shows in the persistent wealth gap between Black and white Americans to this day.

University of Massachusetts Boston economics professor Dr. Dania Francis estimated that between 1910 and 1997, Black ownership of farmland in the South dropped by 90 percent—a compounded value of $326 billion.[29] In *The Black Tax*, author Shawn Rochester cited labor discrimination that cost black Americans $4.4 trillion. Economists Bernadette Chachere and Gerald Udinsky estimated that labor market discrimination between 1929 and 1969 cost Black Americans $1.6 trillion; Rochester revealed through other studies that discrimination in education cost another $2 trillion and housing discrimination cost $1.5 trillion, kneecapping the Black community financially as integration strategies forced further divestment from Black institutions. Dr. Francis also notes systemic racism at the USDA was the biggest reason for Black land loss—although there was no shortage of other pestilence in the form of law, policy, violence, and exclusion that kept weakening Black farmers since well before the USDA came along.

PERCEPTION AND PSYCHOLOGY

Policy has a lot of power. But it goes hand in hand with perception and psychology, the hidden p's of Africulture. Along with the ways that American law made it harder for Black farms to survive financially, more subtle but persuasive activities were at play that also eroded their self-image.

The manipulation of perception of watermelon is a prime example. In the 1910s and 1920s, watermelon was one of the best cash crops for Black farmers. It was easy to grow, it kept well, and you could transport it without damaging it. As an African superfood that was familiar and culturally relevant to Black people, watermelon has been part of the dietary genome for 6,000 years, serving as a great hydrator and a greater nutrition powerhouse; the seeds—which we generally ate a few of—are beneficial for the prostate and the reproductive organs.[30]

A farmer selling watermelons in DC. *Photo by Marjory Collins / Wikimedia Commons.*

But since Reconstruction, when white Southerners first got nervous about independent Black farmers, watermelon has been denigrated as the lowly fruit of the Black community, and both farmer and crop were disrespected and stigmatized. Minstrel shows featured songs about Black people loving watermelons. By the late 1800s, caricatures were appearing on postcards, in movies, in print, and in household goods, showing Black people eating watermelon and looking like childish fools. Some showed Black men with uncontrollable appetites for watermelon, eating it like animals devouring prey, which fed into the other stereotype about their perceived hypersexuality. The 1893 World's Fair in Chicago included a Colored People's Day at which Black visitors were supposed to be given free watermelon—but most of the Black people in the area boycotted the event.

Still, this stereotype kept coming up, and it frequently appeared in pop culture through the middle of the twentieth century and into the twenty-first—and on last Tuesday, depending on where you live. Unfortunately, it hasn't died yet. I wish I could say it's less common now, but the perception and psychological pestilence still lingers to this day in jokes, cartoons, songs, and out-of-touch marketing campaigns approved by individuals with limited or no understanding about the torrid history of this beloved fruit. The vestiges are also still very much heard, every time an ice cream truck pulls into a neighborhood. Many people don't know that an earlier version of the American folk song "Turkey in the Straw" was called "Zip Coon," and it was popularized during the American apartheid era at minstrel and vaudeville performances. American banjo player and recording artist Harry C. Browne (who later became president of the First Church of Christ, Scientist) adapted the song to a far catchier tune called "Nigger Love a Watermelon, Ha! Ha! Ha!" in 1916. In his charming intro, Browne calls out that colored man's ice cream is watermelon. Ice-cream shops played this tune in their shops in the 1920s, and it became the song ice-cream trucks have used for the last sixty-plus years to summon children and adults. As I and many of my contemporaries ran to stand in line to get our orange Creamsicles, Strawberry Éclairs, Bomb Pops, orange Push-Ups, or ice-cream sandwiches, we were serenaded by agricultural bigotry in the key of G major.

The result was the stigmatization of watermelon for the last six generations of Americans. This degradation discouraged African American consumers from buying a crop that had been supporting independent Black farmers across the South. Since a lot of Black farmers relied on Black customers to buy their produce, it hit them hard when this crop became associated with the worst racism of the larger white society. Mothers wanted to protect their children from the evil intent behind caricatures of African American babies with big red watermelon-eating lips that were common in mainstream media, as well as on the walls of stores. These images and the associations they evoked could affect employment, support of a business, access to education, or a litany of other opportunities. Watermelon was associated with all things bad, backward, and unappealing in Black communities for several decades.

Watermelon was a major cash crop for Black farmers. *Photomontage by William H. Martin / Wikimedia Commons.*

By contrast, look at English and American spinach, and the positive, encouraging consumption of its leaves by Popeye the Sailor Man. You think spinach sales were up after Popeye debuted as a cartoon? Spinach and watermelon are both superfoods, but Popeye was a positive marketing tool whereas the watermelon stereotype was intentionally negative. They both targeted children and had a significant impact on adults and their purchasing activities. Choose one of those foods and you might be called a racial slur; you might be made to feel ashamed for eating it, especially in front of a certain demographic of people. Choose the other and you could feel strong, virtually indestructible.

Results of this smear campaign not only took away income from Black farmers, it fed anti-African sentiment. There's a subtle connection between how farmers perceive themselves and what they grow—Black farmers were affected by this stigma. The seedless watermelon was created partly out of this negative perception of the watermelon and its Negro consumer. African Americans were often mocked for spitting out the seeds of these ancient melons, and having a seedless melon made it appear a bit more civilized in the broader American society. The watermelon perception aided in shaping a broader, negative perception of Africa, including calling it "the dark continent." The negative public relations campaign for Africa's image would only get worse as America stigma wasn't through with Africa yet.

In 1985, a Talented Tenth of the who's who of American pop musicians came together to record "We Are the World." The producers and the performers did not have an intent of denigrating Africa, and the song may have raised money for a good cause, but it also cemented negative and incorrect perceptions of a continent of fifty-four nations, turning it into one "unfortunate" country in the American mind. The song also attached a perception that these "poor Africans" couldn't feed themselves. Yes, there was a famine—in one region of Ethiopia between 1983–1985. Ghana also experienced a drought and bush fires in 1982 and 1983 that spurred a famine and pockets of malnutrition in the country. Just like during the Dust Bowl of the 1930s in the United States, economics, political actions, climatic shifts, and other environmental factors created a perfect storm of conditions for a disaster. In the mid-1980s, Americans were bombarded with images of children with bloated stomachs and flies on their face, while listening to America's top pop stars belt out chords encouraging us to help.

As Winston Churchill noted, "Never let a good crisis go to waste." And, indeed, the 1984 famine in Ethiopia was twisted into a destructive negative perception weapon. The American government at that time was interested in creating new markets for American farmers—that is, for white farmers—overseas. That interest often morphed into the official food, security, and political policy of the US government, along with disruption and market volatility in the form of banana republics, regime change, and political upheaval in targeted nations.

This effort began back at the time of the Kissinger Report, a 1972 national security memo that focused on the need to protect American security interests by fomenting unrest in less developed countries.[31] Population control was part of this, but so was food control and market access opportunities for American farmers. Roaming the markets of Cape Coast, Wenchi, Techiman, and Sogakope in Ghana, I witnessed the impact of these policies. Rice sellers had bags of Arkansas and Texas river rice along with jasmine rice brands from Thailand, Malayasia, and other Asian nations. But local Ghanaian rice farmers had a tough time finding markets. Most Ghanaian consumers loathed the idea of purchasing local rice, which was healthier but oftentimes not as polished and processed as the foreign rice. I encountered Ghanaian farmers growing wetland and dryland rice, brown rice, and red rice, who were capable of producing jasmine and other fragrant rice but lacked the milling capacity at a small scale to provide the polished white rice that consumers were looking for. Despite having a much lower cost of production, often the local rice was higher priced than the foreign rice. Foreign governments generally procure products from farmers and sell it to foreign nations for less to grab market share and also to gain political favor and viability with their agricultural constituents. It was eye-opening to see boxes of chicken labeled with American brand names traveling on an eight-hour bus trip from Accra to Wenchi, with no refrigeration, dropped off at a local chicken-selling business along the way. Ghanaian farmers, however, had no markets in the United States for fresh produce or grains because of food safety concerns and issues. Ghanaian pineapples, some of the best and sweetest in the world, are not available at all in the United States. These examples I witnessed in Ghana fit with the US government's goal of cultivating export markets for Americans.[32]

"Starving people" in Africa were a market that could feed two birds with one piece of bread (being a bird lover, I'd rather feed two birds than kill the same two birds with a singular stone). The American government exported their rice, corn, and soybeans along with other commodity crops and US "technical assistance" and support to international markets that we are told aren't able to feed themselves. The crisis in Ethiopia was an opening salvo, the appetizers of a PR campaign that perpetuated the idea of American exceptionalism and saviorism. These actions weren't new, it was the same playbook and strategy used in the so-called banana republics, the Caribbean and Central American nations that grew America's number one fruit, the Cavendish banana, of which very few are grown for wholesale purposes within the United States.

Nationwide campaigns such as "We Are the World" program impressionable minds. It wasn't just entertainment; it was the news. Top mainstream television journalists such as Tom Brokaw, Ted Koppel, Peter Jennings, and Dan Rather were speaking about it, saying "This is what we're doing to support Africa." There was psychological pressure on Black Americans to prove themselves as different than these people who can't feed themselves. At that same time, hip-hop culture was pushing African consciousness in the late '80s.

THE WORD IS BAMAS AND SARDINES

There was a steep decline in the rural Black population between 1900 and 1974—from 77 percent down to just 8 percent. Times and aspects of culture were changing; the youth of the 1970s and '80s were being influenced nationwide by our cousins and family who were living in America's cities. Stevie Wonder's "Living for the City" referenced a taste of the realities, perils, and change of culture that a young man from hard-time Mississippi experienced.

The connection of the plight of Black farmers, Black communities, and even Black businesses and government and social policy comes full circle, because what brought members of my family to cities like DC in the first place was the Great Migration—which itself was driven by assaults on Black farmers and landowners that, consequently, made life harder for the farmers who stayed, because it took away many of their customers. The majority of Black farmers primarily sold to members of their community. If most of your customers are moving away, can you really sustain your business? Convenience became what was sold to residents in urban areas, and it spread

During the Great Migration this family moved from Texas to Chicago two months after two of their sons were lynched. *Photo by* The Chicago Tribune */ Wikimedia Commons.*

throughout the county, and convenience is one of the biggest killers of farming operations.

Because the Great Migration was made up of people looking for a safer place and potential ways to support their families, it grew an idea that the only ones who stayed behind were those who didn't have a desire to grow or be successful, or didn't have a desire for the education that's been deemed a sign of success. When I went to church, mothers and fathers were bragging on their children—"They are doing well in Richmond . . . or DC . . . or Philadelphia." "They got a job in Chicago." That was the goal, to leave these rural areas and be successful. And those who left usually didn't have a plan for going back and reinvesting in the communities that had shaped and molded them.

There wasn't exactly a schism between the more rural Black culture and the urban hip-hop world—but there was no meaningful association from the hip-hop world with agriculture. Farmers were perceived as country bumpkins. Conversations I heard between younger and older people in my community at the time showed that the younger generation was conversing with a new level of proud African consciousness and awareness, touting African history, while the older generation was trying not to rock the boat. "Don't come in here with that African stuff" was a phrase I heard much more often as I got older and wanted to shift conversations to some new African-centered knowledge I had heard or learned. Those images of bloated bellies only validated ignorant perceptions around the continent of Africa.

The DC vernacular around this time developed a term of disrespect and disdain, "bama." "You being a bama right now . . ." "They dressed like some bamas . . ." Bama was a shortened version of Alabama, and it meant anyone from the South.

Despite their differences, urban and rural people still need one another. An urban song in the 1980s that was almost as popular as "We Are the World"

GRAY NEW BALANCE—996 TO 1600 PENNSYLVANIA AVE

Washington, DC, fashion and style in the late 1980s through the '90s was monochromatic—black, white, gray, maybe navy blue, and red. DC was reflective of the beginning of the metropolitan north and reflected the colorways of most northern areas and metropolises up to New York City. Gray New Balance shoes were and are the staple of the city, echoing the drab, lifeless nature of the concrete buildings and the majority of its landscape. You see the same in many urban centers in the American West—a gray or black power suit, with black shoes, a white shirt, and with a power tie as the only pop of color.

But most African Americans in the South and West during this time period wore more colorful clothes, reflecting their environment. Reds, blues, oranges, greens, and pinks were represented much more in the color palettes of fashion in these places where you could see fields of flowers, birds, and green grass. It's reminiscent of West Africa, where the colorful arrays of kente cloth, batik, brocades, and tie-dyes mimic the vibrancy of colors found in nature and the landscape. Black and red, and white and black are reserved for funerals. When I purchased or wore those fabrics in Ghana, the seller or my tailor inquired, "Who died?" It's an amazing phenomenon that those in more tropical environments traditionally wear more tropical colors, and those in cooler, rainier, or concrete-based areas don the colors of cloudy or dreary days and nights.

was "Sardines" by the go-go band Junkyard Band from Washington, DC. Junkyard Band was a group of African American males, as young as thirteen, who performed on pots, pans, hubcaps, crates, and buckets as improvised musical instruments, with the signature go-go sound. They were mostly from Barry Farm Dwellings, a housing project and longtime African American community in Southeast DC, initially established by the Freedmen's Bureau as a settlement of newly freed enslaved Africans. I first saw Junkyard Band perform in the 1983 movie *D.C. Cab*, and one of my aunts took me to see them perform on Nineteenth Street in Northwest DC. Mesmerized by their creativity, their sound, their age, and their rhythms, I became a go-go head.

Go-go was started by the legendary Godfather of Go-Go, Mr. Chuck Brown, in the mid-1970s. The basis of go-go is the percussion, the drums, and the congos (the name for conga drums in the DC area), that keep a continuous beat between songs. It pulls from various cultural sounds, including the Guaguancó conga beat from Latin American music, and an African-style call and response that makes the audience part of the band. It was contagious—my friends and I would play beats on a desk, a book, a school bus seat, using a pencil or a middle finger knuckle. Bands are known for their original songs as well as covers or hooks from other songs.

Junkyard Band recorded two of the most agriculturally and socially relevant songs of the mid-1980s. "Sardines," which described their meals and food insecurity situation after cuts in food stamp allotments, was one of their first hits and is still a go-go classic. "The Word" is a catchy, rhythmic, intelligent song about food and farmer aid being reallocated to defense spending by the Reagan administration. Mr. Brown, the farmer described, loses his crop to bad weather while an anonymous mother is wondering why her Women, Infants, and Children (WIC) assistance was eliminated. The special supplemental food program for women, infants, and children started as a pilot program in 1968 under the USDA and became a permanent initiative in 1972. The Reagan administration repeatedly tried to cut funding to the program, most notably in 1981 and then again in 1985 with hopes of eliminating the program. The propaganda of the welfare queen, first used in a 1976 campaign speech by then Governor Ronald Reagan, described exaggerated claims of a singular person who was accused of welfare fraud in 1974 in Chicago. She became the poster child for why welfare should be cut, exploiting perceptions that Black people were abusing the system, and that a majority of Black people were on welfare. The reality was, and is, that whites made up the lion's share of nutrition assistance then and now. However, in the 1980s, African Americans made up as much as 40 percent of recipients. None of these proposed and actual cuts to nutritional aid programs took into account the social engineering that led these communities to need and depend on the social safety net of federal nutrition assistance programs.[33]

The three-and-a-half-minute song is a brilliant display of rhythm, storytelling, and real-time analysis of government policy, specifically the Farm Bill, which was affecting residents in the nation's capital and around the country. The decades of pejorative policies were a pathogen that had contaminated the soil of community to the point people began to think of them as normal. The extinction event for agriculturalists of African descent was well underway.

THE TRAUMA OF THE LAND

On April 2, 1968, a tragedy befell the Carter family. For three decades, my grandfather and his brothers and sisters had been maintaining the farm that Gran had saved from the clutches of financial malfeasance. The ponderosa, as my uncle Lee Scott called the wooded paradise surrounding the rustic two-level house of his mother-in-law, represented refuge, peace, sanctuary, security, and self-sufficiency, the ability to provide for ourselves and pass something on to future generations. It was the gathering place, the place you could always rely on coming back to, during good times and not so good times.

Lewis Alphonso Carter.

On that cool, somewhat damp spring day, while Uncle Alphonso was riding on an embankment on his Farmall tractor in one of the back fields, the tires slipped and the tractor flipped over. The weight of the tractor on my uncle's neck was too much for his body to sustain. April 2, 1968, would live on in infamy in our family. Uncle Lewis Alphonso Carter made his transition into the ancestral realm, leaving behind a wife, three children, a mother, ten siblings, and a community that came to truly value his soft power and the host of skills that he possessed. He was lost in a moment, and many lives—and the farm—would never be the same.

Two days later, Dr. Martin Luther King Jr. was killed by an assassin's bullets not soon after realizing that he may have integrated his people into a burning house. In a matter of two days, we lost an icon in our family and an icon of the of the Black community.

In my family, the death of my Uncle Alphonso is still felt today. Though he died ten years before I was born, I grew up feeling the vacuum of his presence. An event like that marks the people and the place, at times creating a sense of, "We need to be somewhere else." Many summers I was the only child on a lush 150-acre parcel, playing with the dog, Lady, fighting imaginary enemies with sticks, playing stickball by myself, watching my grandmother or great grandmother cook, or sitting outside shooting the breeze with my grandfather. My cousins and other family my age not being present at the farm during these summers was partly because of the trauma of the land and the severing of the relationship with the land. His death was a seismic catastrophic event in addition to other family members' deaths. Uncle Alphonso's red Farmall tractor, which I pass every time I travel to my farm, is a present reminder both of his presence and our loss that, although I ignored as a younger person, is part of our trauma and pain landscape, like a crater from a meteor shower.

Back in 1968, we did not understand the mental health effects this tragedy would have, quietly rippling throughout the family. Nowadays, there is a greater value placed on professional counseling and there's more encouragement to share how you're feeling, recognizing and respecting grief, as opposed to saying, "Get over it." A quiet code I absorbed from the older males around me implied that crying wasn't strong, showing your pain wasn't strength, and expressing your emotions of grief and bereavement wasn't manly. But consider a plant that doesn't get the moisture it needs. It will wilt, which is an outward sign to the caretaker that the plant needs more moisture. The plant is sharing that something is not okay, that something is not right with me and that I'm not able to function fully until this deficiency or loss (in this case of water) is attended to. Humans wilt in their own way when they're in need, and in truth, it is not weakness to let our wilting show.

The resilience of the land is like the resilience of people. And both can be invisibly depleted by trauma and abuse. The land can be associated with the

Uncle Lewis Alphonso Carter's grave marker.

tragedy and traumas that happen on upon it, and these traumas can subconsciously affect our relationship with the land. A rare few times, I was told to pick a switch, which is a long twig or small tree branch used for corporal punishment—or as it was called in my neck of the woods, spankings, whippings, or beatings. For myself, because of my experience with spankings, switches do not trigger me. But for some of my more rambunctious cousins, I can't say the same, as they had a different, more frequent acquaintance with those twigs and branches. The pestilence our farmers experienced has at times made it hard to see what the land can actually provide—like the reality that it at least gave my family a stable place to live. I had never been to the place where Uncle Alphonso died until I got back from Ghana in 2018. None of my cousins in my generation had been to that part of the farm, never explored the land fully. It became almost forbidden, or an unspoken sacred ground, because of the trauma, loss, and tragedy associated with it.

Uninformed, outdated phrases related to working or being outside, such as "I ain't gonna be out there working like a slave," I've heard all of my life. I've probably thought it a few times myself. Many of these attitudes have a foundation in some sort of personal, family, or historical trauma related to the land. The land is and has been innocent in all of these circumstances, but it doesn't limit and change attitudes and perceptions around it. This is the pestilence of perception that came down on Black families, Black farmers, and many Black people, and it served as a leading driving force to lower our numbers and our presence on and with the land—and almost drove me away from our ancestral lands, heritage, legacy, and responsibility.

RESISTANCE AND THE PIGFORD CASE

The first step toward healing from pestilential experiences is proper information. In my teaching and speaking, I not only offer my students facts about the challenges and inequity Black farmers have faced, I make sure people know that since the time of slavery, Black people in America have consistently, brilliantly, and effectively *resisted* oppression. Just like plants, we have an inborn ability to fight back. We don't always get the credit for it, and resistance is often framed as violent aggression. But the truth is one of innate resilience and creativity.

Resistance continued to flower in political, economic, and cultural ways during Reconstruction, American apartheid, the Great Migration, and beyond. In my own lifetime, there's a standout example of Black farmers finding a way to fight decades-long discrimination: the Pigford case.

In the early 1980s, my mother worked with Mr. Carlton Lewis, a USDA staffer who later became a branch chief and civil rights compliance officer at the USDA. Uncle Carlton became my neighbor (and my first barber), and he often told me that I should go into agriculture. But I didn't know until recently that he had worked, from the institutional side, on the foundations of one of the most famous court cases related to Black farmers and farmers in general in US history.

For decades, the USDA had been routinely denying and delaying loan applications from Black farmers.[34] The USDA itself found in 1994 that "minorities received less than their fair share of USDA money for crop payments, disaster payments, and loans."[35] Black farmers, especially, often found their applications denied on technicalities, lost, or delayed.[36] In most cases, applications and subsequent complaints about discriminatory treatment were thrown into an unmanned office in the basement of the USDA buildings, that is, a trash can. Disaster relief was also routinely denied to African American applicants while white applicants received the support. Most of these loans were processed at the county level, and even in majority African American counties, very few of the committees who made the decisions included Black farmers. This was another major contributor to the decline of Black farmers and the loss of Black-owned land.

Uncle Carlton told me about being asked in the 1980s to investigate a backlog of discrimination complaints. "They put me on a special project and told me I would have so many people working for me," he said. "I said no, I'm going to put a team together with the people I want, and we are going to get this resolved. I put a team together of twenty-five people; I took the best specialists from each agency. The first week we turned over about 200 cases. Seventy percent of them we had done in the first month." USDA staff couldn't believe that his team was doing the work so quickly and so well, Uncle Carlton shared. He was told that the complaints would have to be sent back for further investigation, and that's when they stalled out. "They still messing with the same cases I did twenty years ago," he told me.

His office adjudicated a case that eventually became *Pigford v. Glickman*, a class action suit filed in 1997 over USDA racial discrimination between 1981 and 1996. It was named for Mr. Timothy Pigford, a farmer from Pender County, North Carolina, and Secretary Dan Glickman, the secretary of agriculture at the time. Independent consultants had already found numerous problems with the agency's administration of loans for Black farmers and the way they'd handled complaints in the early 1990s. Many have referred to the USDA as the last plantation because its policies and practices haven't been conducive to sustaining and supporting Black farmers consistently. The denying of operating loans, delaying of loan payments, and limited or no sharing of information about USDA programs, among other issues, has earned this institution, designed to help farmers and promote agriculture, some rather dubious distinctions.

Local officials in his county asked Mr. Pigford, who farmed in Bladen, North Carolina, to testify before the House Judiciary Committee on September 26, 1984. The committee was investigating claims of discrimination by the USDA Farmers Home Administration, which had denied Mr. Pigford loans for his farm over the previous eight to ten years. Unbeknownst to Mr. Pigford and his wife of fifty years, longtime educator and advocate Mrs. Janice Pigford, they would become victims of retaliation and further discrimination for their testimony and subsequent involvement in the largest class action lawsuit at the time against the United States. Around the same time the Junkyard Band was speaking about Farmer Brown and springtime floods, Farmer Pigford lost all of his crops during Hurricane Diana, which devastated the North Carolina coast in the fall of 1984. Everyone seemed to get disaster relief loans, except the Pigfords. When I spoke with Mr. Pigford he told me that this denial was direct retaliation for his testimony. He was told the only way he would ever get another loan from the FHA was if Senator Jesse Helms approved it. Jesse Helms, to say the least, wasn't considered a friend or advocate for anyone of African descent.

Mr. Pigford was encouraged to sue the government for discrimination and to make it a class action lawsuit. He and Mrs. Janice Pigford, Mr. John Boyd, Mr. Carlton Lewis, my statistics professor Dr. Donald McDowell (who helped organize the data to make the case), Mr. Lloyd Wright, Dr. Shirley Sherrod, Mr. Lucious Abrams, Mr. Eddie Slaughter, and other leaders for the claimants did heroic work from a legal, information gathering, farmer organization, and administrative standpoint.

The Pigford case set a precedent for the injustices experienced by Black farmers, even after the plaintiffs won the case. It was decided in 1999 in the US District Court of DC by Judge Paul L. Friedman, who ordered the USDA to make restitutions of more than $2 billion, one of the biggest civil rights settlements in history.

In the mid-1990s, Mr. John Boyd, longtime Black farmer advocate and president of the National Black Farmers Association, was on my MTV news feed way too often, strolling down the streets of Washington on his tractor or his mule named Struggle. It wasn't until later that I made the connection that several of the men who influenced me during my youth were not only teaching and practicing agriculture, they were also fighting for Black farmers at the local, state, and national levels.

The Pigford case provided a landmark ruling, but it was an imperfect solution to say the least. There was pushback from the defense lawyers, and legal maneuvering relegated or diminished the case for these Black farmers. The payments were offered through two tracks. Track A offered a flat payment of $50,000 and was considered an easier way for farmers to prove their cases. Track B offered potentially larger payouts but a bigger burden of proof. Whereas more than 22,000 farmers filed claims under Track A, fewer than 200 filed under Track B.

Two billion dollars sounds like a big payout, and when farmers who aren't Black hear that sum, they make an assumption that "y'all got paid." But Track A still kept people in debt. If a farmer received $50,000, but owed $400,000, they still ended up losing. And a large percentage of the farmers had their claims denied. Many others missed the 180-day deadline to apply for relief.[37]

Because of that, a second round of settlements, called Pigford II, was negotiated in 2010. This resulted in another $1.2 billion for those who had missed the original filing deadline. Again, more than half the claims filed were denied.[38] In the early 2020s, there was a little more movement—a 2021 Equity Commission at the USDA and a $4 billion fund for Black farmers through the 2021 American Rescue Plan was established. But the fund attracted a lawsuit from white farmers, *Strickland vs. USDA*, which got held up in court, and it ultimately never came into fruition. The fund for Black farmers eventually turned into a $2.2 billion program to be administered without regard for race called the Discrimination Financial Assistance Program (DFAP). A federal judge ruled in favor of the plaintiffs in *Strickland vs. USDA* and issued a nationwide injunction that prevented the USDA from using the socially disadvantaged tag for increased payments in the 2022 Emergency Relief Program.[39]

With the USDA policy shifts that have occurred between June and July 2025, I am reminded of a prediction by Virginia farmer Mr. John Boyd in a 2002 ABC News article. Even then, Boyd predicted that farming would be the "first occupation to become extinct for black people."[40] When I started writing this book, I maintained some level of hope that Black farmers would persevere and find a way to triumph. But the ways in which the funding and research to support Black farmers have been attacked and removed concerns me about their future. On July 10, 2025, the USDA issued a memo stating it was ending the unconstitutional preferences based on race and sex in response to the *Strickland vs. USDA* court ruling. The final rule in the federal register also states that "moving forward, USDA will no longer apply race- or sex-based criteria in its decision-making processes, ensuring that its programs are administered in a manner that upholds the principles of meritocracy, fairness, and equal opportunity for all *participants*."[41] This will be problematic for Black farmers, especially when standards of merit, fairness, and equality are meted out by individuals who have a different set of perspectives and life experiences. I know many long-standing USDA staffers who opted to take a buyout in their agency during the Department of Government Efficiency activities in early 2025 rather than to continue working with the USDA, as well as many younger Black staffers who were victims of the last hired, first fired reality at the agency. We do not know what to expect, but many farmers I speak to on a weekly basis are not overly optimistic of the future for Black farmers. Books such as *Dispossession* and *The Shadow of Slavery* by Pete Daniel lay out the history of this discrimination in far more detail than I have space to include here. Daniel's work and that of others illustrate what has happened in the past when merit, fairness, and separate but equal were used as determining criteria for Black farmers. As I've said in this book and in talks I've given around the country, we are experiencing a Black farmer extinction-level event, and it appears as if time has sped up to bring this to pass in my lifetime.

It's discouraging to me that thirty years after the Pigford decision, I am still doing work similar to what Uncle Carlton and Mr. Pigford were doing when I was in primary and elementary school. I expect my children and grandchildren may have an even tougher fight in this arena than I have. From the time of Mr. Pigford's class action case, from before that and still to this day, there's pushback to the progress that we've made through our resistance. There's pestilence in this attitude: "Y'all didn't deserve this; you're exaggerating these things."

It's like tillage—when you keep disturbing the soil, a plant has to devote its energy to constant rebuilding as opposed to the innovation and work it's really good at. My goal is for my children not to take on this aspect of my work. If they want to be farmers, they should just be farmers and not be Black farmers. This is the plight of the African, the burden of carrying melanin, our version of chlorophyll.* In America, there's a high cost to being kissed by the sun.

I sometimes try to sing my own version of the 1977 classic song called "Zoom," recorded by The Commodores. My version sums up my Black farming past, present, and future and the val-YOU I place on farmers.

I am a foolish dreamer,
but I don't care,
however my happiness I'm not sure about.
I'm searching for a silver lining,
horizons of justice, fairness, understanding,
and equity that I have never seen,

---

* According to nutritional biochemistry researcher Arturo Solis Herrera, one of the key properties of melanin is that it creates energy by disassociating light from water, which then helps glucose provide usable energy for human beings in the same way that chlorophyll works for plants to create energy.

but it appears like that may be all just a dream that I dream.
I tried to fly away to Africa,
where my mind could be fresh and clear,
where everybody can be what they wanna be,
without fear of retaliation, stigma, retribution, poverty, or trauma.
I/We would like to know the taste of honey,
the sweetness of our efforts,
for once in our life.
I'd like to fly away . . .
we've been wanting to fly away
from this persistent pestilence for the last five centuries,
so we can be what we wanna be,
seekers of happiness
waiting out there somewhere.

W. E. B. Du Bois would likely be unsurprised to know that the problem of the twenty-first century is still the problem of the color line, though our country has doubled down on the pretense that there is no problem. James Baldwin, if he was living today, would undoubtedly still say that America changes all the time, without ever changing at all.

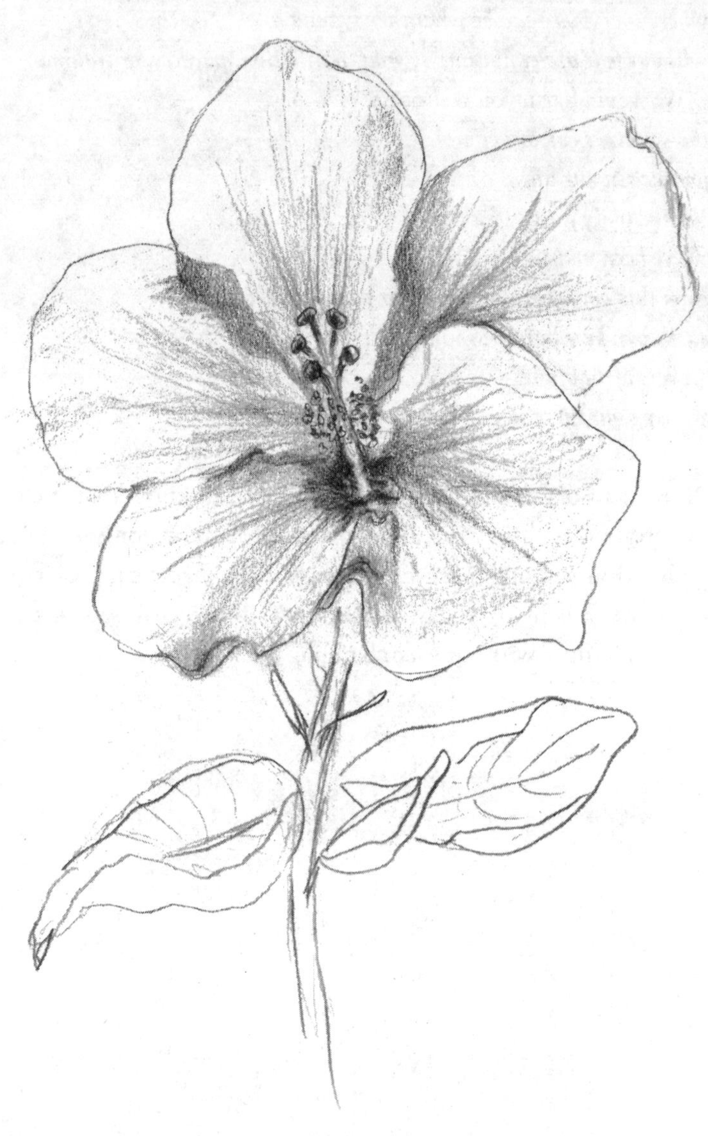

CHAPTER 8

FLOWERING

ALLY, SAVIOR, OR ACCOMPLICE?

Just bees and thangs and flowers . . .
Everybody loves the sunshine.
Roy Ayers

Every plant has its proper time to flower, and you can't rush it or wait on it. For example, when the window of time opens for tomatoes to make fruit, they first create flowers. These flowers must experience ideal conditions. If nighttime temperatures are too hot, the flowers will abort and drop off, and there will be no fruit. The vanilla orchid, whose story I told in chapter 5, is another example of a plant that need specific conditions for pollination.

Flowers also need the help of other organisms to be pollinated: insects, birds, bats, the wind, and even humans. In the case of the fig tree, only specialized fig wasps can properly pollinate the flowers. When pollination is successful, the fruit will be abundant. Social change is a flowering process, too. And in human terms, the pollination collaborators are people who want to help others. If that pollination happens correctly, in the right window of time, the fruit will be lasting results, lasting change.

Chances are fleeting in America for leaders to raise challenges that change minds or social norms, especially when there is no immediate profit motive attached. There was an opportune time, a flowering time, between early 2020 when Mr. Ahmaud Arbery, Ms. Breonna Taylor, and Mr. George Floyd were senselessly murdered, and around June 2023 when the presidential election cycle kicked off. That flowering time was a chance for significant change to be made, for justice to be meted out or at least addressed financially as a fitting response to past, historic injustices. "Strike while the iron is hot" was the message percolating through my planning circles for those three years. But what we saw once again was that creating change requires both swift, decisive action and long-term commitment. When we start looking at liability or politics to determine how cautious we need to be, that's when the fruit is aborted and the flowers of change fall to the ground.

We witnessed all of this in the aftermath of Mr. George Floyd's demise. People changed because the cell phone–recorded lynching of an American citizen brought forth deep emotions, not felt for generations, not since the time when innocent Black men were being lynched at ungodly rates. In 2020, comedians, entertainers, musicians, athletes, and the average person were moved to speak out about some of these gross systemic realities, but it was nothing new. Ms. Billie Holiday was singing about strange fruit; Mrs. Nina Simone was goddamning Mississippi, which made Mr. Marvin Gaye want to holler and throw up both his hands. Mr. Otis Redding and Ms. Aretha Franklin were asking that you give us a little R-E-S-P-E-C-T and Mr. James Brown had to reiterate that we were Black and proud, and damn right, we are somebody. Mr. Richard Pryor, Baba Dick Gregory, Mr. Redd Foxx, Mr. Chris Rock, Mr. Dave Chappelle, and many other comics made realistic jokes about harsh realities of our communities. However, when all of this historical trauma bubbled up in late May 2020, we saw many performative, profit-preserving actions instituted very swiftly, from corporate DEI programs to the renaming of institutions, the removal of monuments and statues, and the establishment of Juneteenth as a national holiday. Even the performative passing of the 2022 Emmett Till Antilynching Act, about 130 years late, and over 65 years after Mr. Till's unfortunate demise. What was missing was a long-term commitment to change.

Now, a few years later, we're seeing the backlash. A lot of what was built after 2020 is being taken apart, being strangled financially, socially, and politically. Black people aren't surprised. Many of us who are students of history were waiting for this heel turn, this realignment with whiteness over rightness. As organizations, businesses, and institutions were writing these new mission statements, sidebar conversations focused on taking advantage of this opportunity while it was here, because it wouldn't be here long. We know America, how she moves, responds, reacts, and retaliates. Black people have consistently taken a higher road, trying to work with the system. And I compare it to when you see flowers that have opened on the tomato plant and you get excited, not realizing that the conditions later in the night will destroy those flowers. You don't understand why this keeps happening, year after year, decade after decade, administration after administration.

In some ways, 2025 feels like we're back in the 1920s, with federal initiatives, programs, and policies being rolled back. On January 29, 2025, the forty-seventh president of the United States executed an executive order that paused and cut funding for various federal grants and agriculture-related programs within the USDA. I felt like a Jedi (which, ironically, is also what certain equity initiatives were called) in *Star Wars: Episode III – Revenge of the Sith*, when Chancellor Palpatine / Darth Sidious issues Order 66 to stormtroopers to destroy the Jedi order. Memorandum M-25-13 from the Office of Management and Budget declaring the funding pause was a devasting blow to small farms, climate-conscious agricultural research educational institutions, and the regenerative agricultural industry. The season of flowering was over, and we don't know when we'll get another chance to make fruit.

But this chapter is about solutions—the same ones I've been talking about since before George Floyd was killed, the same ones I intend to keep talking about. Solutions aren't timeless; they do have an expiration date. However, they need to be implemented anyway. Racism is white peoples' invention, so who can solve it? Only white people and those adjacent to the benefits of identifying as white in this society can. At the same time, the reality is that the mindset that created the problem can't solve it alone. To address this centuries-old problem will require the minds, thoughts, emotions, and realities of its victims, with full compensation for their time and energy in providing solutions. Solutions are only as good as their implementation, as was proven with the Civil Rights Act of 1866. It becomes a matter of will to implement whatever solutions are out there. The solutions I propose here are useful tools, but only if applied consistently, and tweaked as necessary.

As I write this, it bothers me that I had hope, that I thought that America could be something other than America, that if I could do the requisite work, put in the required time and energy, then my children wouldn't have to deal with this. It was my fervent thought that my children could be the first generation of farmers in my family to just be farmers, not a sixth generation of Black farmers facing the same challenges, discrimination, and adversity that their great-great-great-grandparents faced. If American perceptions don't shift, I know that my sons are going to be a target—as all young Black males will be. If they are going to be a target, I want them to be a target for success, a target for growth, a target for support for their entrepreneurial endeavors. The other type of target Black males become is part of the America that I despise and tried to escape. Because of my legacy and what I feel is my purpose for this season, I was unsuccessful in my escape attempt. But recent events have reminded me of my naivete in thinking I could be an instrument of change in a society bent on never changing to accommodate my community's best interests.

"Make America Great Again" does not make sense from an African American standpoint. What era are you referring to in which America was great for my community? The Ronald Reagan era, the Lyndon Johnson era, the Jim Crow era? There's never been a golden age for African Americans in America. If we had been left alone to tend to our traumas, grow our communities without interference, have equal justice under the law, build self-sufficient and self-sustaining communities, and become independent, there could have been one, but white America repeatedly prevented this from happening. As. J. Edgar Hoover once memorably phrased it in an FBI memo, the US government spent much of the 20th century actively interfering with Black nationalist movements to prevent the rise of a Black "messiah."[1] This has served and continues to serve as a source of trauma and pain for many who have tried to make the land of their birth their home.

THE BIRTH OF AFRICULTURE

Agriculture is critical because it's a microcosm of America. To know the heart of America, you can look at the soul and the soil of its farms; its character will be revealed. Being a farmer without subsidies, crop insurance, price supports, and guaranteed markets is a tough road to travel. Being a Black farmer makes the road seem more like something out of *The Odyssey*, a perilous journey where you don't know what obstacles you will face or if you will survive. But you keep sailing, because it's in your blood.

I first began experiencing the realities of the American agricultural system after I returned to the United States from Ghana. Selling African vegetables to restaurants intrigued me as a business niche, and I also had a job at the Virginia State University Small Farm Outreach Program as a coordinator for the Small Farm Resource Center.

A colleague of mine recommended that I apply to a leadership program called Virginia Agricultural Leaders Obtaining Results (VALOR). She had been in the program and said it would be great for my development. I didn't think I needed much more development, but I saw the value in networking opportunities for continuing to build out the resource center at VSU.

In this program we networked with industry professionals at various levels of the agriculture industry in Virginia. One weekend a month we visited agricultural operations all over the state, and one week each year we took a national and international trip—like we were the agricultural national guard.

Not long after our sessions began, I noticed various forms of implicit bias, racial illiteracy, and cultural incompetence within the agriculture industry. The industry as presently formulated is largely a monocrop, both in Virginia and throughout the country. A monocrop is a large field or acreage planted with just one species of crop. In terms of farmers themselves, the monocrop is people of European descent. You might already expect that the majority of farmers are white, but it's way out of proportion to the demographics of Virginia or the country. Black people constitute eighteen percent of the Virginia population, but Black farmers make up less than 3 percent of farmers in the commonwealth.

VALOR was my first experience since high school of being in the minority in a group of white people. It took me back to the time I spent in the FFA, feeling like I couldn't find a home in those places. In the course of my adult life, I'd attended college at an HBCU, lived in all-Black neighborhoods in DC, and then lived in Ghana. As I began to associate with present and up-and-coming leaders in agriculture, I noticed that I was also a minority almost everywhere we went. It wasn't surprising, but rather eye-opening; I had to acclimate to the experience.

I also saw more clearly the opportunities I had missed while growing up. At the state fair, for example, I saw young people showing their cattle and receiving up to $4,000 apiece for their animals. That's the kind of experience that encourages, supports, and incentivizes the next generation—it proves that there's money and career-sustaining opportunities in this industry. As a teenager, I had no clue, even living with an FFA advisor, that that was possible. If you had told me in 1994 that I could keep a cow in the field and earn that kind of money, instead of working at Kings Dominion, the local amusement park, I'd have considered livestock rearing as a potential part-time if not full-time summer job. But I had never seen money in agriculture. Witnessing these activities now, from my more mature vantage point, motivated me to want to share them with the next generation.

Vestiges of our agricultural past are still alive in the agricultural industry, sometimes in painful ways. To visit a cotton gin and see pictures on the walls of Black people picking and ginning cotton was a stern reminder of our storied and exploitative relationship with that crop. During a group task of picking cotton seeds out of a fresh cotton boll, I thought about the fifty or more years that the hands of enslaved Africans were the cotton gin for antebellum America. The stain of past racial exploitation was smeared on many of the agricultural settings we visited in my twenty months of journeys around the state with the program.

The experience wasn't all negative, especially in the southwest and northern parts of the state. The zenith of misunderstanding occurred toward the end of my sojourn with the program. I should have known something was wrong when, as we were driving to an orchard deep in the heart of the Piedmont region, my phone signal disappeared, right as we passed a small trailer home adorned with the battle flag of the Confederacy. Walking into the orchard's country store, I was shocked but not surprised to see pictures sometimes erroneously referred to as Black Americana displayed on the walls. Metal signs of Uncle Remus and watermelon-eating girls in plaits, known disrespectfully as pickaninnies, welcomed one into this fine establishment. (See page 7 in the color insert.) I'm sure it was owned and patronized by good people on both sides.

At every VALOR session, I consistently tried to explain my viewpoint calmly and build understanding with my colleagues, but I felt like my words were not getting through to those who needed to hear them the most. Taking time away from my family to go to these sessions became less appealing. I started to sour on this networking experience. In January 2019, we met for a discussion with the then governor of Virginia, Ralph Northam, along with some of his staff from the Virginia Department of Agriculture and Consumer Services. The following week, he was a leading headline on the cable news networks because of photos discovered in a mid-'80s college yearbook of Northam in blackface, looking like a Jim Crow version of Thomas Rice in his heyday. Let me clarify: These were photos from the 1980s, not the 1880s. Photos from an era when a Black man, Jesse Jackson, was running for president, and L. Douglas Wilder was a rising star in the politics of the commonwealth.

Despite my efforts, the group never directly discussed racial realities in Virginia agriculture. I would point things out but then the conversation would dissipate. These issues were family issues for me; I was researching how my own family story connected to these experiences I was having in the program. Throughout the program we were expected to write blog posts on the program website. In my posts I started off with a lot of optimism—I wrote in my first dispatch, in October 2018, "Interactive and slightly spicy sessions are indicative that this may be an interesting two years. I look forward to respectful discussions, debates, and throwing in my curve balls and sliders into the discourse. . . . My legacy in agriculture precedes the establishment of the Commonwealth of Virginia and the Constitutional Convention. It is only right and responsible that I gave honor and recognition to my founding mothers and fathers who built this nation and authored my family book."

By February 2019, after the Northam scandal broke, it seemed that my perspective was an issue for some of the other participants and my blog posts were sounding less optimistic. "I had every, and I mean every, intention on not really mentioning or addressing racial issues with our last VALOR expedition in January. . . . The Commonwealth has a problem, it's a prejudiced, biased, racist and white supremacist culture and rule of law. . . . My gorgeous flawless golden caramel honey mocha skin tone can be costly to my immortality on a regular basis, because of general perceptions the country and the Commonwealth has of large black men. Arriving safely at home after our Richmond VALOR session is a privilege not a right for some of us."

In addition to site visits, class discussions, and writing assignments, participants were expected to create a project or program that would further our work. My experiences spurred a wish to create a program for those of African descent who needed to see and experience agriculture in a different light, with a different perspective, and without a biased undercurrent. If that undercurrent were present, the program participants would address it and discuss it fully. I wanted an agricultural program and experience that would be relatable to people of African descent, to instill a sense of pride in our agricultural accomplishments, and to share opportunities not readily available to our community. After almost every VALOR session, this vision of an uplifting and purposeful program was percolating in my mind as I drove home.

VALOR was a compost experience for me—something that I didn't enjoy, that was a bit stinky and uncomfortable, that I had to grow through. I was kicked out of the program at the end because I missed one or two more sessions than their threshold allowed. It was not a surprising end to a less than comfortable experience.

The impetus to create an alternative to 4H and FFA for Black youth grew as I shared my presentations and workshops with middle and high school students around the state, planting seeds and seeing a spark of interest or even a bit of appreciation in some of the young people for whom the traditional FFA or 4H groups didn't fit or appeal culturally.

Through my job at VSU, I had also been leading focus groups with farmers. Many of them said the same thing: "We can grow anything, but we can't always sell it." That was an impetus to find markets for my fellow farmers. I also conducted workshops at VSU about ethnic vegetables, and I would hear farmers, as well as people from different cultures, say, "Oh, I know this one." Talking about these crops became a constant educational process for me as well as a way to connect with other cultures. I was seeing common threads and pondering African crops that seemed to be superfoods, although they weren't usually talked about that way. This expansive subject opened up so many doors for learning, increasing cultural literacy and understanding, and providing market opportunities for farmers. I thought, "How can I grow this into something more?"

The idea to start not only a business but a nonprofit made practical sense. It would allow us to better provide technical support and catered services for both new and experienced small, "historically underserved and socially disadvantaged" (as the USDA once described farmers who weren't white men or military veterans) farmers I worked with. In addition to supporting farmers, a nonprofit would allow me to teach about ethnic vegetables, get Black youth interested in farming, and spread the message about African contributions to agriculture. The organization could serve as pollinators for growth and flowering throughout the sphere that holds Black farmers, African foods, and the history of Black people on American farms and farms around the world. The seeds of Africulture had been in me for a long time, but I didn't know what exactly would grow from them.

BLOOMING CONNECTIONS

Once Africulture was properly established as a nonprofit, it started to grow like a Nigerian spinach plant. A small amount of resources went a long way in a short time. I presented at Virginia Tech, Virginia State University, and

AN ANCESTRAL RESPONSIBILITY

The idea of bringing other people along is an ancestral responsibility and a seasonal purpose for me. Historically, my family has always looked to help other people once we got on our feet. When my fifth great-grandfather Mr. Robert Ellis helped found Freetown, other families joined him to assist in building their community. It was the same thing with my great-great-grandparents who purchased the land on Carters Lane, then deeded off lands they owned to help other relatives get started. Even my great-grandmother gave up one acre on the edge of her property to an elder or sister she went to church with, to make sure they had a home. I've witnessed my mother give of herself and her resources for those in need to this day, and I've likewise witnessed my father's service through agricultural education. This has grown to be part of my ancestral and biological résumé—another part of myself I could never escape.

youth seminars for 4H. As I developed and deepened these presentations, I realized they were not just for children, college students, or small farmers. This was important information for those industry leaders and staff like the ones I worked with in VALOR.

One of the positive aspects of my VALOR participation was that I had met and networked with many industry professionals. Because of my questions and my presence (I was the only Black man in most cases, and often dressed in tailor-made African print clothing), I stood out. The secretary of agriculture knew me, and commissioner of agriculture Dr. Jewel Bronaugh out of VSU gave me a lot of positive encouragement. (She later became the first Black deputy secretary of the USDA.)

In my research for Africulture, I started to become an advocate for equal funding for VSU in line with the Evans-Allen Act of 1977, as well as equal funding for 1862 and 1890 land-grant schools generally. I had discovered how underfunded our small farmers program was, and I was quite confident that this was one of the few budget line items that had not increased for twenty-five years. I presented information about this lack of funding to whoever would listen at VSU.

I saw this as a potential time for flowering. Virginia had a Democratic governor with a major faux pas involving blackface, a Democratic-leaning state House and Senate, a commissioner of agriculture from our university—this was the time to ask for $1 or $2 million. The plant was ready to bear fruit, but what happened to the flower? It aborted, dropped off. The budget at VSU remains the same through 2026, at $394,000 annually to service approximately 35,000 small farmers. (Do the math: That's less than $12 per farmer.)

Meanwhile in my Africulture work, interest was increasing exponentially. The connections and, more important, the relationships were just as pertinent as the information. After I gave a talk, participants often told me, "I didn't know that" or "I never thought of it like that." I've spoken in metaphors since before I knew what a metaphor was, and found myself trying to create the metaphors that could connect everything under the umbrella of this wide-ranging topic, a peek into how my brain works, creating a Power Rangers or Voltron-like structure I called Africulture. I wanted to highlight that Black people are very much a part of planet Earth and the interconnected universe, and I wanted to use the Africulture platform to promote connections on many different levels. The African ideal that everything is connected always resonated with me, a stark contrast to the siloed, isolated, subjective thinking in most of the West. The connections among music, food, soil, agriculture, literature, meta- and quantum physics, nutrition, and mental health maybe aren't normally seen, but they are made clear to me through my Africultural lens.

Africulture became an official nonprofit in 2020. Small farmers found themselves in a unique situation that spring because of COVID-19. Small farmers across the country became a lifesaving force for families and communities, setting up farmers markets and driveway stands where customers could come and pick up produce, proteins, and value-added products during a time when entering a grocery store felt dangerous. It showed the flexibility of small farmers to manage transitions quickly in response to changing conditions, and also how vital we can be for national security.

Then in May 2020, Mr. George Floyd was murdered on every television, cell phone, tablet, and laptop screen in America. This tragedy greatly heightened the racial conversation throughout the country. Calls and emails to Africulture from media outlets and institutions around the region grew five-fold as consumers, organizations, institutions, and governments tried to figure out how to support the marginalized communities of Black and other minoritized farmers. My initial cynical thought was, "Where y'all been for the last twenty or thirty years?" But I also wanted to take advantage of this time for flowering when it appeared.

Realities and practicalities was my prime focus, sharing insight from a small farmer perspective. But I knew that it was about more than media stories. Before growers could plant anything for these potential new buyers, they would need assurance that there would be access to markets throughout their growing season.

One of the most significant institutions who committed to change in 2020 was the University of Virginia. I met the head chef at UVA back in 2018, when a fellow farmer and retired physicist, Mr. Clarence Rodwell of Llewdor Gardens, was trying to pitch them on buying his microgreens. We learned then that they serve approximately 90,000 meals a week to a very diverse student body. In 2021, I was invited by Ms. Shantell Bingham to assist in helping UVA with achieving their sustainability goals and objectives with socially disadvantaged and local farmers. We had an initial meeting with UVA Dine, an Aramark company, and explained what we wanted: a reach-down approach to helping us out, as opposed to us having to reach up to their standards.

We made this ask not because our farmers couldn't grow good crops. The issue was the official certifications that institutional buyers including Aramark usually require of growers—such as Good Agricultural Practices (GAP) certification. This certification program allows retailers to trace their sources if produce has to be recalled, and it allows for accounting of food safety practices to keep retailers from getting sued over things like *E. coli* outbreaks. It's expensive to earn certification; at the time only two Black farmers in Virginia were GAP certified: Willie Mae Farms and Goldman Farms. And only one of those, Goldman Farms, grew a wide variety of produce (Willie Mae specializes in sweet potatoes). It sounds cute to buy from local "farmers of color" (not a term I use, but one I may regurgitate when it's used in my presence), but I reminded them that our options were limited and there were realities that we, supplier and buyer, had to work through together.

Cost was of critical importance. How much were they going to pay for our farmers' produce and protein? It was important to me and my farmers to receive a higher price for our produce because of the demand for products and to make up for not supporting farmers in the past. Our discussion with Aramark and UVA (which included 4P Foods, a food hub and food distributor) revolved around this price policy.

After an initial meeting we were all supposed to appear on a panel together, along with Mr. Tom McDougall from 4P Foods. But the Aramark officials backed out. I was frustrated and irritated, but not surprised: When the tough questions were presented, it appeared that Aramark executives decided this cooperative effort wasn't right for them. However, six months later, they contacted us and said they were ready to talk.

I was shocked by their reappearance. It turned out they had been doing some serious reflection, and through internal deliberations they had come up with a practical solution that didn't jeopardize or compromise their food safety standards but still met our needs. They had taken the time to honestly figure out what they could and most importantly couldn't do, and they communicated this to us directly. It was a great lesson in institutional responsibility. An endeavor like this requires time and patience. Aramark is such a large institution, and they needed time to figure out how to adapt their complex system of soliciting and procuring produce and proteins from local growers. I hadn't been as patient as those executives were, but because they committed to change, it ended up being a meaningful initiative.

On a farm tour in our region, we invited Aramark chefs and staff to my farm and two other farms, and we introduced some history and the realities that small farmers and chefs face. After the tour Dr. Paul Freedman, a professor at UVA, offered me an opportunity to share this information more widely through teaching a course at the University of Virginia.

I began teaching Africulture as a course in UVA's environmental thought and practice program in 2022. UVA once was a place where my grandfather couldn't even have walked around on the campus unless he was carrying a broom and dustpan. His life wouldn't have been allowed to flower there at all, but my message is finding an outlet because of Dr. Freedman's and others' understanding of the need for students to become more well-rounded in learning about this story and their roles in it. For me, to be able to teach my family story and the reality of what Black farmers have faced and what they have influenced and been influenced by, at a university level, is significant. Planting seeds in my students and having them work on projects to address solutions for this industry has been inspiring. Students Mr. Kanaya Jones, Mr. Jordan Murphy, Mr. Garreth Bartholomew, and Mr. Carter Purves have continued to work and plan with me after their time in my classroom. In the first three years of the program, two more Black farmers became GAP certified. That was a 100 percent increase and six or seven additional farmers are working on the certification process at the time of this writing. First-year revenue flow from 4P Foods to local Black farmers was less than $20,000, and in 2025, it has hit $500,000. Sales to 4P Foods have increased; they are now purchasing from more than fourteen Black farms and institutions. Not all of this spend was from UVA, but through their intentional support of Black farmers, other minoritized, beginner, veteran, and small farmers throughout the region benefited and gained more access to markets and revenue.

THINKING BEYOND THE MONOCROP

The work of Africulture to date has only touched the tip of the tip of the proverbial iceberg. We look forward to others adding to this field in every manner, culturally, agriculturally, historically, artistically, psychologically, and environmentally to expand the purview and the connections among the myriad cultures that may be labeled as Black.

The plant world depends on a great variety of collaborators for pollination. Flowers and pollinators experience mutual benefits. Herbalist Stephen Buhner writes that plants have the ability to change the composition of their nectar or pollen to attract diverse pollinators. The more different forms life can take, create, and thrive, the more robust and energetic the whole system is able to be.

In the human world, Western conventional agriculture is based on dominating the farm ecosystem and explicitly working against diversity of both plants and people. Varieties of apples grown in North America at one time numbered hundreds or thousands of types; now, many grocery stores sell only three or four shelf-stable varieties. Thousands of species of edible, nutritious crops are largely ignored, while America's farms churn out a few monocrops—corn, soybeans, sugar beets, rapeseed, wheat, and rice—to provide the vast majority of calories that Americans rely on for food. The monoculture permeates every aspect of industrial-scale agriculture from the crop inputs to the genetically modified seeds. This capitalistic and mechanistic approach is a growing problem and is not beneficial for human health or the health of planet Earth, so why does this system persist?

Most likely, it's the profit motive: Monocropping concentrates the revenue in the hands of large farms, which account for a mere 22 percent of all farms. The other 88 percent of farms are small family farms, and they produce just 19 percent of the value of farm products sold.[2]

The solution to the problems that monocropping creates is to promote biodiversity instead—and that includes *the people who grow the food.* Western conventional agriculture in America has produced a monocrop of white farmers. Farm ownership is one of the whitest occupations in America, and that's striking considering that at one time farmwork was *primarily* accomplished by Black labor and farmers. In the same way biodiversity benefits the soil and the land, it benefits the economy and the community. We have allowed that to be systematically stripped away.

If we're going to rebuild a more diverse farming industry, we need all types of pollinators and collaborators, working their magic with all types of flowers. I see myself as a pollinator for seed companies, farmers, urban farm managers, and many other flowers in the gardens and farms of Black agriculture, some of whom have cross-pollinated me, as well. If we don't protect and pollinate flowers, they either wither or they get picked. It's vital to continue pushing back against the losses of Black farmers and landowners, and to support everything that Black farmers can bring to the larger community in terms of culture and nutrition.

THANKING THE POLLINATORS

As a young flower, coming up I was pollinated by so many amazing bees and wasps as well as the wind. My father, Mr. Michael Carter Sr., was my first and has been my most consistent pollinator, literally and figuratively. My godfather, Mr. James Gibson, was also an agricultural teacher. Uncle Gib's father, Dr. David Gibson, and his lovely wife, Mrs. Elsie Gibson, kept me actively engaged outside whenever I visited their home in the warmer months. The wisdom and steadfastness of Mr. James G. White, my other agriculture teacher in high school, inspired my decision to attend North Carolina A&T State University, where he had been a student. Surprisingly or not, the deacons at my church, specifically Mr. Floyd Bates, his brother

Brother Act Wins Soil Contest

A brother act which was assisted by a first cousin gave Orange County High School's FFA soils judging team an area championship earlier this month. Warren Carter, kneeling left, his brother Michael, center, and first cousin, George, right, were joined by classmates Bill Feldman, standing left, and Ray Breeden, standing right, in an area Future Farmers of America competition. Soil Conservationist James Hale set up the contest which pitted high school students on their ability to distinguish different types of soils. Warren Carter closed out the event with high point honors and a score of 135. Cousin George was close at his heels with a count of 134, and the Orange team was victorious with a top mark of 397. Madison County High School came in second with 326 points. The local team is shown with a display board of various Orange County soils. Members also won a trip to the FFA regional contest at Virginia State College.

My father, in the center, won an FFA soil-identifying contest in 1971 along with his brother Warren (*left*) and their cousin George (*right*).

Mr. Woodrow Bates, and Mr. A. B. Howard served as pollinators, demonstrating agricultural hard work and true definitions of manhood. Dr. Maury Grainger, Mr. Jody Martin, Dr. Donald McDowell, and Dr. Alton Thompson pollinated my college flowers and assisted in bearing the fruit of my agricultural economics diploma. In the mid-2000s, I heard Dr. Kweku Andoh, an elder from Ghana, lecture at Everlasting Life in Maryland on Ghanaian healing herbs. Prince Yadiel ben Israel along with my spiritual brothers, such as Ahtur Dahveed Jawara and Ahk Yahneev Israel, actively grew my spiritual fruits and understandings related to the metaphysical nature of agriculture.

Baba Tarik Oduno, a towering agricultural figure in the Washington, DC, area and in urban agriculture, who stood about six feet, ten inches tall, pollinated young adults everywhere he went. Baba Oduno was an active and engaged elder I met back in 2000 at a United Negro Improvement Association meeting in Northwest Washington, DC, on the second floor of the Reeves Center. His catchphrase was, "There's no culture without agriculture." I probably didn't truly fully digest this phrase until I returned back to the farm after my sojourn in Ghana. Baba Oduno planted seeds of agriculture everywhere he went or was seen. Like a towering oak, he was an imposing but gentle presence who influenced hundreds if not thousands with his presence and words.

My dear brother Mr. Renard "Azibo" Turner, who has a great agricultural and philosophical mind, has continuously pollinated my perception and understanding. Azibo has been growing in Louisa, Virginia, for about thirty years, and he is an example of steadfastness, focus, and diligence. He and his lovely wife, Mrs. Chinette, have been through so many challenges as farmers, but are still at it steady and strong. I first met them in 2007, and our bond was immediately cemented. Visiting his farm, I could see his ingenuity and tenacity in every aspect of his operation. Their Akbash, a breed of Turkish livestock guardian dogs, kept all mammals on the farm (including myself) in order and protected.

Azibo also kept goats; he created his own breed called Bangus. I was amazed at his brilliance, but as he shared with me, markets for his produce and proteins weren't there despite the stellar quality. As manager of a restaurant at the time, I wanted to make sure we were actively supporting and patronizing a Black farmer for our produce needs. But our menu was designed more for supplies of produce from a conventional distributor. It was a challenge to use Azibo's purple carrots for our orange carrot supreme salad or lacinato kale for our garlicky kale salad designed to use blue scotch curly kale. This experience, however, assisted me greatly in working with chefs and restaurants in the future, understanding how to fit local produce into an established system.

Azibo and I have stayed connected; he provided the produce for the catering at my wedding and I have taken several groups of young people to his farm to see this example of agrarian intelligence and will.

Azibo stayed in communication during my sojourn in Ghana. While I was there he became a poster child for small farmers, serving on various committees and boards for industry organizations and the USDA.

After I returned from Ghana, I had no idea that he was about to get foreclosed on because of a USDA loan gone bad. He wasn't able to pay large balloon payments and by July 2018, his land was going up for auction. I was devastated, for him and his wife, and for the larger agriculture community. My great-uncle and cousin met me at the auction; I was there to offer emotional support, as the $479 in my savings account didn't give me much confidence to bid. My cousin stayed in the thick of the bidding, down to the final two, but a phone bidder kept out-bidding him. We were unsuccessful in acquiring the property, though successful in supporting our family friends when they needed it.

This story is an example of what can be lost when there isn't proper support for a farmer who is thriving and flowering, but still vulnerable to the discriminatory treatment of Black farmers in the United States. If there had been better pollination for Brother Azibo throughout his farming career, he may not have ended up at the point where his entire operation was at risk.

As the wind blows over our fields and pastures at Carter Farms, a family member will say that the wind is Uncle Tiny, which was the family nickname of my Uncle Alphonso. And indeed, wind is a consistent and reliable pollinator. *Ruakh*, the Hebrew word for wind, is also the word for spirit. This definition resonates with me, especially as I recall a quote from the Honorable Marcus Mosiah Garvey's 1924 speech in Harlem: "Look for me in the whirlwind or a storm, look for me all around you, for with God's grace I shall come back with countless millions of Black men and women who have died in America, those who have died in the West Indies and those who have died in Africa to aid You in the fight for liberty, freedom and life."[3]

Though I never physically met Uncle Alphonso, I feel him in the whirlwind that crosses the farm on a daily basis. The ruakh, those divine winds that blow over the land, are the spirits of those who came before us. They pollinate regardless of our awareness of their presence. As his breeze runs over my skin, my breath becomes unified with his presence, and I'm reminded of all the ancestors who tilled the soil, in freedom and enslavement, whose energies form and are blown through these winds, and through my lungs, breathing life into me.

Pollination is somewhat mysterious. We see insects fly from flower to flower, but we cannot observe the process within the flower itself that results in the production of seeds. We only know that flowers and insects collaborate in a rhythm and a cycle greater than themselves. And we know they both benefit from that relationship. Pollination means doing it for yourself and for others. It's different than picking flowers and taking them away for your own enjoyment. Most flowers will soon perish after being plucked, without producing seeds.

TRUE POLLINATORS

As a culture, Americans have a tendency to pick more flowers than we pollinate—to be more worried about protecting our own needs than helping others. There is a need for commitment to real and lasting pollination. What does that look like?

During that wellspring of interest in Africulture in the early 2020s, I would pose a question to people and institutions who invited me to lead classes and workshops: Are you *interested* in change, or are you *committed* to change?

Sometimes a social movement seems like the right thing to do because everybody is doing it. People write mission statements and attend one-time events. It feels good in that moment to do something charitable to help the less fortunate without threatening your overall wealth or standing. That's what I call being *interested* in change. Being *committed* to change is sticking to the change even when there's pushback, even if you are threatened with funding cuts because of your DEI program. Commitment means risking personal finances, status, and occupation to stand for what is right and correct despite the repercussions. Commitment doesn't care about getting credit for being brave; commitment has no desire to be a savior. Commitment is much more about what you do than what you say.

There's a scenario I've seen repeatedly. When specific racial matters start to affect resources or opportunities, or the work becomes a bit more uncomfortable or unpopular, individuals or institutions may lose their energy and back away from commitments that have been made. Those in leadership will tell me they don't have the power to bring about change, but it was individuals who set up the discriminatory situation in the first place, and policies from the past can always be undone. I don't have a choice about being committed because I don't want my children to face the same realities I'm facing today.

Changes in levels of commitment are especially apparent in times of conflict. President Abraham Lincoln was interested in change, but he was not fully committed to change; he was ultimately more concerned about keeping the Union together than ending slavery. His secretary of state, William Seward, was committed to change. Seward, a cosigner of the Emancipation Proclamation, was a staunch abolitionist who was also very good at understanding the psyche of his electorate and foreign diplomacy. Seward utilized his understanding of the electorate and chose not to reveal the Emancipation Proclamation until after a perceived victory, the Battle of Antietam, so the Union would not appear to be asking for help from enslaved Blacks in the Confederate states. Instead, it would appear that the enslaved were asking for the government's help. (Lincoln had wanted to issue the legislation after consecutive Union defeats on the battlefield.) After Lincoln was killed, his successor, Andrew Johnson, showed that he was not even interested in change. Johnson wanted to protect the status quo—he broke down Reconstruction and did his best to take the country back to its prewar state. That tendency to protect the status quo is evident again now in the political repression rising up in the United States.

A lesser-known but interesting figure is General William Mahone, a railroad magnate, slave owner, and Confederate general, who nonetheless became a considerable force for the advancement of Black people after the war was over. He became a Virginia senator and founded and led the Readjuster Party in Virginia, which briefly controlled the state government and enacted progressive policies—supporting Black suffrage and citizenship, uniting the interests of poor Blacks and poor whites, lowering taxes, and reinvigorating education. Mahone ended up branded a race traitor by other white leaders and isn't widely remembered today.[4]

William Mahone. *Photo by Magnus Manske / Wikimedia Commons.*

His legacy does live on in the form of Virginia State University (originally named the Virginia Normal and Collegiate Institute), an 1890 land grant university that he helped to found. Mahone was a person who put his money where his mouth was—he committed himself to change. And unlike Howard University, Spelman College, Lincoln University, or Morehouse College, VSU is not named after the white philanthropist or leader who funded or founded it. It's nobody's trophy. To me, Mahone is an example of how to use resources and money to do the right thing for those who were once his enemies, even if the conflict had ended just a few decades earlier and the wounds were still healing. Using money and influence to effect change without promoting yourself demonstrates your commitment to change. Mahone's life directly affects me; I wouldn't exist without VSU.

Maybe Mahone was influenced by his experiences with brave, courageous, and bold Black troops and leaders in the war. Once you've been exposed to Black excellence, when you see the value Black people bring, your respect for them changes. You may begin to appreciate the value of people who just need a fair shot in order to reach their potential.

Value is one of the most critical components and principles that Africulture seeks to embody. Princeton University professor and author Dr. Eddie Glaude speaks often of the value gap, the way advantage and disadvantage are distributed according to the valuation of particular people, generally how one's racial grouping is perceived and valued. The value gap, or value tax, is a notion that some people should be valued more or less because of the color of their skin, which affects the distribution of advantages or disadvantages to those people.

I don't know how much Mr. Mahone valued the formerly enslaved, especially during the war. However, the newly freed men and women involved in his political institution were given advantage to run for and hold political positions, many times winning elections and representing the interests of their communities. At times there was a backlash, tremendous backlash. In 1883, in Danville, Virginia, because of the success of the Readjuster Party in winning control of the Danville city council, white supremacists took to violence and terror in an attempt to brutally wrench back political control of the city. Five deaths of African Americans resulted. Political freedom for Black people was a very real threat to white supremacy in the Danville area.

One of the Readjuster Party's active and leading members was Delegate Alfred Harris, a lawyer and farmer. A fierce debater in the Virginia General

MEMBERS OF THE GENERAL ASSEMBLY
1887-88
Front Row—Left to Right: Alfred W. Harris (Dinwiddie), William W. Evans (Petersburg), Caesar Perkins (Buckingham).
Back Row—Left to Right: John H. Robinson (Elizabeth City), Goodman Brown (Surry), Nathaniel M. Griggs (Prince Edward), William H. Ash (Nottoway), Briton Baskerville, Jr. (Mecklenburg).

Alfred Harris, lower left, and other Virginia General Assembly members. *Photo by Pbrks / Wikimedia Commons.*

Assembly at the time, he authored the bill that would create the institution that later became Virginia State University. (Harris is another great man with whom I share a birthday—Leos are pretty amazing.) He served as the first treasurer and first secretary of the school's board of visitors. He, as well as fellow legislators and farmers Delegate William Faulcon and Delegate Amos Dodson, successfully introduced legislation to start the Colored Agricultural and Industrial Association of Virginia in January 1886, which established an annual state fair to display, exhibit, and market agricultural and industrial products. Several other agricultural associations sprouted up across Virginia and the South at this time, seeking to galvanize populist whites and Blacks toward better prices for their agricultural products and collective buying of supplies and goods. These turned out to be very short-lived organizations and associations, made irrelevant by the more common racialized separation of the Black Codes and punitive consequences for whites who worked with Black farmers and communities.[5]

Sawubona Readjuster Party. Sawubona Senator William Mahone. Sawubona Delegate Alfred Harris. Sawubona Delegate Amos Dodson. Sawubona Delegate William Faulcon.

The impact of the Readjuster Party proved to be beneficial for at least a decade, until the racialized politics of Virginia eroded the party and some of its gains in terms of policy and representation at the local, municipal, state, and federal levels.

One of the signers of the Declaration of Independence was Philadelphian Benjamin Rush, who's considered the father of American psychiatry and the founder of American medicine. He is the literal face of American psychiatry, his likeness being the logo of the American Psychiatric Association. He was also a charter member of the Philadelphia Society for Promoting Agriculture and a strong proponent of agriculture and agricultural innovation. His advocacy supported the idea that agriculture is central to public health because it assists in promoting mental health as well as physical health and well-being. He didn't believe, like many in his time, that mental illness was caused by demonic possession; rather he saw it as a disease of the mind.

Mr. Rush advocated for the abolition of slavery; he was considered the leading and loudest voice against the enslavement of Africans, arguing that slavery had a detrimental effect on the minds of the enslaved. In the *American Journal of Psychiatry* in 1970, Betty L. Plummer's essay "Benjamin Rush and the Negro" explains that Rush cites hypochondriasis as one ailment caused by being enslaved. In today's terminology, this would be referred to as culture shock. He also felt that songs and dances were an enslaved person's symptoms of insanity, madness, grief, and melancholy. A coping mechanism is how we'd describe it, but it was also part of who the enslaved were—parts of their original cultures.

Rush also promulgated the idea that being Black was a disease similar to leprosy, and that he was going to uncover the cause of this malady. Rush's premise was that if the Negro's blackness could be removed, it would provide a level of contentment and happiness and achieve a oneness of the human race. His solution to the race problem was to make the white citizenry comfortable by removing the melanin-infused skin, making Blacks more acceptable to whites.

Rush's support of the Philadelphia Free African Society—especially finding financial support for the African Methodist Episcopal Church, founded by Minister Richard Allen, and the African Episcopal Church of St. Thomas, directed by Reverend Absalom Jones—would prove highly instrumental to the physical freedom movement and activities of the mid-1800s, including those of the Loyal League. Rush found partnership in the free Black community on many occasions. He used the virology understanding and medical support of Reverend Allen and his community to assist in controlling a yellow fever outbreak in Philadelphia in 1793, and he assisted young brilliant medical minds such as Mr. James Durham to have their genius heard through his own vocal cords, when their own voices couldn't or wouldn't be heard because of the color of their skin. Sawubona Reverend Absalom Jones. Sawubona Minister Richard Allen. Sawubona Mr. James Durham.

In our time, people who want to be true pollinators need to become real accomplices of Black farmers, rather than allies or saviors. Allies are only interested in change while accomplices are committed. Saviors want to be included in the credits for creating change without following the leadership or interest of the people. Accomplices are in it with you, and risk taking the fall or the punishment if caught. Accomplices aren't in it for the credit, they are in it because it's what they believe in and sincerely desire to do. They are willing to risk sacrificing something, or everything, like Mahone did by creating a political party that assisted in empowering African Americans, as well as himself, and declining to put his name on an institution whose purpose would be the education of African Americans.

Dr. Martin Luther King Jr. put it this way: "The ultimate measure of a man is not where he stands in moments of comfort and convenience, but where he stands at times of challenge and controversy."[6]

A SHARED DEFINITION OF RACIAL EQUITY

In order to work toward racial equity, we need to agree on what racial equity means. Equity should not just refer to perceived treatment. Equity should include discussions and ideas around shared ownership and control of resources. If the concept of race encompasses a race—as in a competition—we can understand it as a contest for resources, with skin color determining who is on which team.

In almost every region around the world, European colonizers identified the resources that would be of value to them and figured out the means to extract or capitalize upon those resources. The fight over resources—land, money, jobs, business and investment opportunities, government contracts, grants, time, and so on—continues today. Black farmers and landowners have learned from experience that as soon as a white person wants their land, in many cases, the land will be lost. Mobs would run families and sometimes whole communities off the land, such as in Wilmington, North Carolina, in 1898. What happened to Brother Azibo is another example. The gentrification that happened in Washington, DC, while I lived there, is yet another. Whether it's favorable proximity to the city and the mass transportation system, or the desire for mineral rights or water rights, some aspect of land and ownership is generally the central motivation for the taking of land.

From a business or financial perspective, equity is defined as the value of an asset that an owner possesses after subtracting associated debts and liabilities. On a social level, equity is defined as justice, fairness, and impartiality. The sacrifices made by generations of our African and Indigenous ancestors, and the value of our contributions to this country as land stewards and custodians, entitle their descendants to utilize the first definition of equity. Equity implies ownership, and ownership inherently includes the rights of justice, equality, and fairness as long as that value gap or tax is not applied.

The concept of "home equity" is not generally based on fairness or justice, but rather the value of a home minus how much is owed on it. The home's value is geographically and subjectively based on who lives in the home, where the home is located—but overall the equity is still the equity. It's a store of value for the home. Racial equity is a store of value for people, aligned with human equity. Racial equity has to be about ownership in communities and being able to control or share resources, in addition to justice and fairness. It means asking questions such as: Do you value my life, or my son's life, as much as yours? Would you be upset if my son got the job your son or daughter applied for? Would you be bothered or upset if your children had to work for my children as employees? Will you commit to the rightness of racial equity even when it affects you or your family in a way you may not like?

It astonishes me that people in America and many other Western countries generally view dogs through a color-blind lens, but make immediate presumptions and adopt prejudices simply because of another human's skin tone. No matter their coat color, hair length, shedding capacity, shape, size, breed, sex, or temperament, all dogs are generally seen as equals, their lives respected and their value upheld. It's only against other humans that people tend to discriminate related to biological, geographical, and genetic characteristics.

Racial equity for farmers means the ability to own land and manage it as we see fit, according to zoning codes, not subject to alternative interpretations of these codes. It means equal access to the land and government resources that white farmers have been benefitting from for hundreds of years. As I'm writing this in 2025, however, some of the racial equity solutions we had been working with have now been made illegal at the federal level.

LEAN INTO DISCOMFORT

Black people have always been forced to placate white people so they can stay in their comfort zone. It's a rare occasion when a white person willingly says, "Let's have a conversation and get this out of the way."

And thus, Black people don't say certain things to white people; we don't bring up certain histories. When we speak up, we can risk being seen as the aggressive Black person and losing our lives, our jobs, being kicked out of school, being put into jail.

The act of teaching diverse perspectives on American history and culture is now experiencing backlash in schools in 2025, sometimes couched as a concern that children should not be made to feel uncomfortable. But the children whose comfort is prioritized are white children; no consideration is being given to the perspectives of children who look more like me. When I hear statements like this, I think about Grandma Lucille and Grandma Rose and how uncomfortable they were when white children disrespected them by calling them by their first names. Yet they had to suffer through it. There was no consideration of their comfort, their feelings, or their dignity.

Diversity, equity, and inclusion (DEI), like race or racism, is not a creation or invention of the Black community or Black masses. It did not prioritize or focus solely or primarily on Black Americans as racism historically has. White women and other minoritized groups are far larger benefactors of DEI programs in the long run, with more limited and targeted inclusion of African Americans.

If you're feeling uncomfortable as you read this book, do some self-reflection about why. What have you been taught that makes you uncomfortable with the new information or scenarios I am presenting? Increasing racial literacy is often uncomfortable. As psychologist Carl Jung postulated, the shadow is the unconscious, unaware part of our personalities that we do not readily see about ourselves. Doing shadow work involves addressing, confronting, dealing with, embracing, admitting to, and resolving the shadow. America has a shadow that precedes its existence as a society and nation—it's the shadow of racism that extends back to the fourteenth century—and yet America overall has refused to effectively address it in any meaningful way.

I encourage everyone to become racially literate. Literacy is defined as being educated or cultured; it is also defined as the ability to read and write. Most people on planet Earth are literate according to the second definition. English has become the standard for literacy in most contexts, but there are hundreds of other written languages that use other alphabets and scripts. The majority of the world is illiterate in any of these, just like one can be highly educated in Euro-American culture but totally ignorant of cultures outside that sphere.

One of my goals with Africulture and this book is to help people improve their racial literacy around a broader history of agriculture. My hope is that learning this history, accepting the challenge of being uncomfortable, will lead people to look for tangible ways to support Black farmers, local farmers, small farmers, and environmentally conscious farmers.

BUY FROM BLACK FARMERS

The most fundamental way to support Black farmers is to buy their products!

I sometimes receive calls and messages from people who live in municipalities 50 or even 500 miles away from me, asking to buy my products because they've heard me give a talk or they've read an article about me. I always advise them to buy from their *local* Black farmers.

Visit your local farmers market and support the Black farmers who are there. Personally, I make it a point when I go to farmers markets to try to buy something from everybody there. I understand from selling at that type of event how much it costs to be there and the effort it takes. My microeconomics professor, Dr. Maury Granger, taught me that we "vote with our dollars." If you want a farmer to thrive, you must keep voting for them to be there at the farmers market. If you don't, they may not be there next year.

No Black vendors at your farmers market? Support the Latino vendors. And inquire with the management. Speaking to management about the lack of certain demographics at farmers markets can encourage managers to seek out more diverse vendors. Consumers dictate what institutions want, and consumers have the power to demand and enhance diversity.

This is the water or fertilizer that grows farmers and ultimately grows the next generation. I've seen Mr. Elisha Barnes, a fourth-generation peanut

THE BLACK FARMERS INDEX

Visiting the Black Farmers Index (blackfarmersindex.com), a directory founded by Dr. Kaia Shivers—presently the most comprehensive nationwide directory of Black farmers in the country—is a great way to begin the process of connecting with Black farmers in your area. The website is divided into regions and is an excellent resource for those looking to locate and support Black farmers and ranchers. Brother Azibo Turner and I were commissioned to start a Virginia Black Farmers Directory (virginiablackfarmerdirectory.com) in 2020, sponsored by Virginia Foodshed Capital (now Foodshed Capital) and the now defunct Local Food Hub. Ms. Brianna Stephenson was hired to take it over after we shared an initial list of farmers, and she did an exceptional job of growing it to include Black farmers and ranchers all over the commonwealth. You can also do a search for Black farmers in a given county; when you find them, call them, visit them, or send them an email or text.

farmer on Pop Son Farm in Southampton County, Virginia, with his granddaughter at a farmers market selling watermelon for $3 a slice. He can cut ten slices from one melon, and that's a nice profit on a single watermelon. But he's also had scenarios where he's had to let a crop die on the vine because he didn't have markets. Which one of those scenarios do you think his granddaughter finds more inspiring, if she's thinking of becoming a farmer herself? If a consumer asks their farmers market manager to bring in somebody like Mr. Barnes, and then buys watermelon slices from him, that is a meaningful act of support and pollination for him and for the future of his family farm.

It takes intentionality to support the people who want to grow food for you. We've discriminated against these farmers for generations; now they need direct support they can count on for the long term. That includes support from Black consumers in addition to white, Latino, South American, Asian, African, and Caribbean individuals, families, and institutions.

This is why Africulture was created, to fight the perception of inferiority around slavery and agriculture, which affects the value we place on food itself. Americans need to redefine their perception of healthy food, too, to stop relying on the convenience of processed food from modern-day grocery stores. People need to be willing to pay more for fresh unprocessed produce, grains, and proteins and to develop relationships with the growers who produce that food. Highlight those individuals and be patient with them as they grow their businesses. Small farmers can't always offer you triple-washed precut salad mix. What they can do is give you some beautiful lettuce grown with love, though they may depend on you to chop it yourself.

As far as people in a position to procure or purchase fresh produce or proteins for an institution like a corporate cafeteria, school, or a hospital, when requests for proposals (RFPs) are created, there should be a percentage allotted for the procurement of foods from minoritized, local, and small farmers. This type of proposal forces the vendors to look for local, small, Black, and/or minoritized farmers. Even a small percentage matters. Including the language "5, 10, 25 percent for local underserved producers" in your RFP can have a resounding impact on the larger local agricultural community.

SHOW FARMERS THEIR VALUE

Valuing food and the people who grow it should be a given—they're literally keeping us alive. But many cultures, including the United States, fail to do this. Ghana is a great example of a culture that values food and farmers: They have a national farmers day, a public holiday, on the first Friday (*Fiada*) in December. There are competitions and rewards to honor farmers, and "farmer of the year" prizes in multiple categories in every region of the nation. Schools and banks close for the day, and awardees are recognized on national television. The Ghanaian public understands how much cocoa and other crops support their economy and how many people work in that industry.

Zambia observes their national farmers day on the first Monday in August. Tanzania has *Nane Nane* ("eight eight" in Swahili), which celebrates the country's farmers on August 8. Bolivia, Peru, India, Chile, Afghanistan, Vietnam, and South Korea also celebrate a farmers day. In these countries, the occupation is valued and respected and provides a sense of pride in what people eat, how they eat, and who grows the food.

Surprisingly, as I researched for this book, I discovered that America has a National Farmer's Day, too. It's October 12. I'll probably celebrate it for the first time this year. (Farmers really need a new marketing team, because this special day is not built into my Google calendar.)

State and county fairs celebrate agriculture, but they're almost always held in rural white areas. Therefore, who wins those competitions? Who gets that encouragement? How are the fairs touted in local, state, and regional media? Farmers, and especially Black farmers, in America do not always feel valued by society. When I first spoke out in public about Carter Farms, I didn't know how to speak with confidence and pride about the value of what I was doing. I was just talking about how I grow crops and sharing (somewhat ashamedly in my eyes) family anecdotes. It wasn't until I led tours on the farm itself or offered a workshop and received positive and encouraging feedback that I started to understand the significance of sharing the stories and experiences of Black farmers. My perception of myself was increased and enhanced by those who came to the farm to visit or help out. Farmers, teachers, students, engineers, accountants, researchers, consultants, retirees, military veterans, doctors, surgeons, politicians, chefs, and many others from around the country and world have traveled to visit my farm and hear the Carter family story, and it's been quite humbling and encouraging. It's the kind of support that can keep the candle lit in the current and next generation of farmers.

There's exciting potential for events on farms that can benefit both the community and the farmer. Think of farmland as a place for education and for healing. I encourage food businesses as well as mental, physical, and spiritual health practitioners to seek the use of farms for retreats, where they can push for greater cultural competence and awareness through conversations around food and food cultures. Farms are places for nature-related therapy that can benefit neurodivergent people as well as those who need a break from urban life. If you're having a group meeting, do it on a farm. Let farmers know that they're the first doctors, that they're stewarding a piece of your healing, and that helps them to heal, too.

HIRE MORE BLACK EXTENSION AGENTS

Representation matters. Extension agents represent the profession of agriculture as a whole in their communities. Extension agents are supposed to know about support programs and share information about them with their constituents. And it makes a difference when that person offering you information and support looks like you—it's some assurance that the person will have your best interests at heart. If almost all agents are white, that carries a message about who belongs in the vocation of farmer.

Agriculture is one of the most subsidized industries in the United States. Given that, if you are part of a group that doesn't receive government support, you won't feel like a full citizen.

The reality, though, is that extension agents are usually stretched thin, and they often focus on the biggest growers, even though the vast majority of all farmers are small farmers. In my state, we need to empower the program assistants with the small farmers' outreach program the same way we empower extension agents. The differentiation between program assistants and an extension agent is rooted in the Smith-Lever Act of 1914, and the separate and unequal "Negro extension service," whose first director was Dr. Thomas Campbell, a Carver protégé from Tuskegee University and the nation's first extension agent. After the Civil Rights Act of 1964, the Negro extension service was phased out, and the commonwealth of Virginia hired a few Black extension agents, like my Uncle Clif Slade in Surry County, Virginia. They served very well in their roles. However, the workloads increased along with the education requirements (in Virginia, the job now requires a master's degree, though the pay starts at around $50,000 a year in 2025). There is a high level of turnover because of this and other factors. Program assistants do similar work in terms of outreach, but they are paid less than the extension agents, generally aren't full time, and have no health benefits. More funding for extension as a whole is needed to keep our farms diverse and growing. Many African-descended federal agricultural staffers left their positions because of policy shifts, staff reductions, and buyouts in the federal government, which could pose problems in the next few years as the agricultural industry moves toward adopting more artificial intelligence platforms.

The average age of a farmer in Virginia is fifty-nine, but for Black farmers it's sixty-three. Between 2017 and 2022, Virginia lost 10 percent of its farmers and 20 percent of its Black farmers. The state is losing Black producers at a higher rate than any other demographic. We have to pass these farms on, and Black agents could help with that process. Culturally diverse agents could reach out to engage younger farmers and help transition farms from one generation to the next. We lose a lot of farms in auctions on the courthouse steps, because some farmers they think nobody's interested in continuing their work, or they pass away without having written a will to address what should happen to their land.

SHARE HISTORY WHERE YOU CAN

I've devoted a lot of time on Ancestry (ancestry.com) to try to learn more about my lineage. Sometimes pop-up messages list new connections to ancestors, but those connections are always to white people or to Europe. My research has taken me back to 1598 with some of my ancestry in England and Scotland. But any ancestry without European links is much harder to trace. For many of my ancestors who were Indigenous and/or African, enslaved or free, who escaped or avoided being raped or assaulted by colonial citizens, I can't find any connections earlier than the late 1800s.

I know very little about my mother's side of the family because outside of her white great-grandmother, everything else we can trace about her family dries up after my great-great-grandparents, between the mid- to late-1800s. My white grandmother's line, however, can be traced back to the early 1700s.

There is a deficit of recorded information about the lineage of Black Americans because of the way information was systematically erased and excluded for several hundred years. But that history is critical for us. When we can find it, it helps us piece together answers and put a greater value on the land we have or hold, since we generally sacrificed a lot to get it.

If you're a white person who has an old family Bible in your possession, or family items in your attic, or access to the local historical society, you might be able to help supply some missing information to Black families. When you share an interest in someone else's history, it helps them see the value of it.

WILL LAND TO BLACK FARMERS

Shawn Rochester, author of *The Black Tax*, uses the PhD initialism to stand for Purchase, Hire, and Deposit. These words sum up a way to aid in restoring Black communities in America from historical and systemic discrimination. Purchase from Black businesses and institutions. Hire African American personnel, contractors, realtors, accountants, etc. Deposit into Black-owned financial institutions.

It's a sound strategy that can be expanded upon, in what I would call choosing to pursue a "postdoc."

Land is an intrinsic store of value, one of the largest vessels of wealth for many. It's also a piggy bank for generational wealth. The world over, land increases in value over time. The land purchased by my great-great-grandparents was sold for $722.05. Today that sum wouldn't cover the annual taxes on the property. Land as an appreciable asset is a great way to pass on the proper equity to those denied it for many years and generations. This could take the form of individuals or organizations gifting or transferring land to African American or Indigenous individuals and/or institutions. Or it could be creating or funding a foundation or trust with a Black family or individuals, using the land as part of the financial base. Share the wealth.

I'm not saying that you should not pass on an inheritance to your offspring or your family. I'm asking that you consider *also* giving to the descendants of those whom you or your ancestors may have taken from in the past. A strategic financial contribution administered through your last will and testament could be a meaningful way to attempt to right the wrongs of ancestors who profited from wrongs and refused to do right. Make a decision to be committed to change in death. Who will argue with you? Your will has to be honored and respected, and your choice could change the trajectory of other legacies and impacts in very positive ways.

SUPPORT BLACK LEADERSHIP

The abolitionist John Brown is an example of somebody who wanted to improve Black people's circumstances but couldn't let go of an image of himself as savior. The leadership council of the Loyal League referred to Mr. Brown on numerous occasions as "Major Brown," to distinguish him from the generals—the Black leaders such as Dr. John S. Rock, Mrs. Mariah Stewart, Mrs. Anna Murray Douglass, Dr. Martin Delany, Mr. Frederick Douglass, and Mrs. Harriet Tubman. Generals don't take orders from majors. But John Brown could not follow instruction; he had to be the leader. He wasn't actually *committed to change* even though he gave his life for his cause. He was committed to being a savior.

I've seen this happen numerous times with well-meaning institutions that dictate through money or opinions, looking to achieve their goals or their own vision of what the goals of Black institutions should be. On a positive note, more recently, I've also witnessed organizations providing resources or funding to Black-led organizations and allowing leadership to be provided by Black organizations including Africulture, Dreaming out Loud in Washington, DC, Happily Natural Day, the Black Family Land Trust, Southeastern African American Farmers Organic Network (SAAFON), OurSpace World, Good Earth Therapy, Hampton Roads Urban Agriculture, and the Persimmon Collective.

Learning to take a back seat, especially after providing finances or funding can be tough, but it is necessary to build and restore trust on all sides. No one wants to lose anything, and for Black organizations like Africulture, we've come into this industry in a financial deficit.

There is much still to be done to reach the level of the equity of ownership. With ownership comes justice and fairness. The land, this Earth, is quite large enough for every living being to have a share. Moving past financial gain and into wholeness will assist in more flowering, and ultimately in consistent, sweet fruits. I'll speak for the choir: We are tired and done with the strange fruits that have continued to grow on the trees of justice in America season after season.

CHAPTER 9

FRUIT

THE FRUITS OF THE ANCESTORS, LEGACY, AND THE FUTURE

My foundation with who I am, reigns supreme
in all aspects of my life. I really am who I say I am.
Tobe Nwigwe

When you are deciding on next steps, next jobs, next careers, further
education, you should rather find purpose than a job or a career.
Chadwick Boseman

I have come to the end of the life cycle for plants, and Black farmers, beginning with the seed in the soil through the first growth of a root and young leaves through the fullness of mature leaves and the sweet potential of flowers. Black farmers in America have persisted despite being planted in the worst kind of soils, and we've held on to as many of our roots as we could despite that contamination and lack of sustenance, while also fighting against waves of cultural, social, political, and economic virulence. My work is to highlight all those stages of our existence, connect us back with the roots of our beloved African soil and crops, help out with pollination of the Black farming community that still exists around the world, and revive communities that are on the verge of extinction.

Ultimately, plant growth and human development lead to the fruit. The beauty of the botanical realm is that fruit is both the fulfillment of the process and the container of the seeds that will start the cycle over again. It's the profound truth of generational wisdom, the genesis and culmination of legacy, and potential springboard for the future. The fruit is a mature ovary. Its primary function is to protect and disperse the seeds, and it does that through its qualities of color, taste, aroma, composition, and nutrition. The fruit is a culmination of the soil, seeds, germination, first leaves, true leaves, and flowering as well as a testament to overcoming the pestilence and challenges that the plant faced.

Fruiting is a responsibility and a survival instinct for most plants. All plants are unique in how they approach seeding and fruiting. Not every plant creates a fruit; some form seeds that are not enclosed inside a fruit. Environmental challenges such as heat or cold stress; pest, insect, and nematode damage; or blight can all prevent fruiting or cause inferior fruit. Some plants are adaptable, yet others may require near-perfect conditions to set a good crop of fruit.

The fruits in our human farming community are the producers, consumers, activists, historians, researchers, content creators, and purpose-driven educators who are thriving at different levels and coming to fulfillment, realizing their own potential through the act of nurturing those coming after them. The fruits are the elders, the youth, and all in between. They may be sweet or tart, suited to eating raw or cooked. Historically, many of our fruits in America have been bitter, producing out of responsibility in extremely stressed conditions, making the next generation of seeds more resilient in spite of toxic and horrific growing conditions. We've been mislabeled, mischaracterized, and misunderstood, because to understand the fruit fully, it's necessary to understand the soil, seed, and rootstock that are that fruit's foundation, along with the adverse conditions it has had to overcome. Over the last 500 years, Black farming elders have maintained their role and function as fruit in dispersing seeds, philosophies, resistance genes, visions, and ideas.

The issues interfering with the fruiting of the Black farming community have remained the same consistently for the last 125 to 406 years or more. The fruit has become what Langston Hughes referred to as a dream deferred in his poem "Harlem." For too many Black farmers, the fruit withered on the vine or suffered flower abortion, preventing fruit altogether. Others had an abundance of fruit just sit in the fields and rot, creating its own problems.

In 2025 there's not as much fruit in the farming community as we'd hoped to nurture, because of the toxic and purpose-harming elements that have affected the seeds over the last five or six generations. Within the agricultural community, farmers of most races and nationalities are facing an extinction-level event. This means that now, more than ever, it is important to express gratitude for the fruits we do have.

ELDERS

When I think of fruit, I think first of the elders who helped to nurture and push me when I was young. I was raised in a family and social environment that was rooted in agriculture since before my debut in this realm. There are so many elders who also played a role in keeping me in agriculture, no matter how frequently I ran away from it. I think of the watering and seed planting that my agricultural elders showered on me, from Mr. Carlton Lewis to Brother Azibo Turner, from Baba Tarik Oduno to Ms. Ira Wallace and the work she continues to do around seed saving and now planting seeds in a new generation of seed keepers. The sacrifices of my elders like Mr. and Mrs. Timothy and Janice Pigford, fighting to get justice for Black farmers around the country. I am grateful to Mr. Eddie Slaughter, who fought until his death for justice for farmers, a double amputee veteran, who almost lost everything in these fights, including his legs. Unknowing of their kinship, I worked very closely with Mr. Slaughter's younger brother, Ahk Yahneev Israel, in the United States, Kenya, Israel, and Ghana for over two decades, just growing and cultivating community and farmers. It's important to me to acknowledge those who came before me, not letting their names or influence to fall to the ground.

My community of elder trees, as well as peer trees and young saplings coming up under the shadows of the forest canopy, have all added to my soil, enriching, strengthening, and growing me. These spirits are eternal spirits in my eyes, many of them still living while others have transitioned. If we think of those elders as the pioneer trees, or the grandfather trees, we can respect them for the way they feed and nurture all the others. In this chapter I want you—yes, you—to acknowledge those elders who poured into you. Write down their names, reach out and contact them if they are still living. It's pertinent to keep those energies and spirits alive in our minds and hearts. Numerous cultures share a belief that individuals die two deaths, the first being a physical loss of breath, consciousness, and being when your heart no longer beats. The next and final death is when your name is no longer remembered or spoken, and your existence is forgotten. A lot of names are in this book, whose spirits and energies I intentionally want to keep in my consciousness and briefly in yours as you read about them. Take time as well to admire and acknowledge the fruit that inspires you, encourages you, and creates a competitive energy for you, or that you desire to emulate or follow in certain respects. It's customary in many food cultures to acknowledge when a fruit you consume is sweet. Some even talk to the fruit before they consume it, as a form of a blessing, a prayer of gratitude and thanks, to convince the fruit to be sweet. There is a need to communicate with that which adds to our lives.

My elders poured and planted into me without the luxury of popularity, with limited or no media attention, and without the dopamine-laden Likes of social media platforms. They did it in the face of bias and discrimination and less than advantageous revenue streams. It's a phenomenal accomplishment that their fire was never quenched, their passion and purpose never overpowered by the prejudice and obstacles they unfairly had to face. They are my heroes, heroines, elders, stateswomen and -men, griots, linguists, historians, culture savers and preservers, and a continued inspiration—an example of what fruit should be. They are my ruakh, my wing and spirit, the proverbial wind beneath my wings.

Competition doesn't exist in the forest like it does in the human world. Trees, understory plants, and mosses share resources and work as complements to those who can manage in the environment of the forest. So one thing I won't do is compete with brothers and sisters of any generation who are actively engaged in this world of farming, gardening, or just growing in general. If I can, I will feed them and help them to grow. Some have grown higher and taller than me. I love them for that because it gives me something to shoot for. And I'm also inspiration for the trees who didn't get a chance to sprout as quickly as I have, or who haven't had a five-generation head start. We can all still support one another, feed each other, advise and counsel one another, empathize with each other and thrive.

These are my soil sisters and soil brothers. When they ask me if I'm available to help, I say yes—I will make this phone call, give you this resource. And other people say the same to me when I reach out: They say, "Whatever you need." The forest doesn't benefit if one tree hoards the resources. We may have individual accolades and accomplishments, but we're not bigger than the forest, and we never can be. The following is an introduction to a handful of my soil brothers and soil sisters who I commune with seasonally. They represent just a fraction of the stories, experiences, and devotion to agriculture that I use for my own inspiration and motivation.

MR. CLIF SLADE

I call Mr. Clifton Slade "Uncle Clif," adding the title like I do with most men of his era. Uncle Clif and his wife, Aunt Arnetta, have been friends of my parents, and now of my family, for decades. As a third-generation farmer at Slade Farms in Surry County, Virginia, he's a familial uncle and a valued mentor and innovator. He and his cousin Mr. Glenn Slade have been giving me advice not just as a farmer but as a man since I was a teen, and I respect their voices and try to apply their example to my life and farm.

Slade Farms specializes in sweet potatoes and elephant garlic—especially the production of sweet potato slips. Sweet potato slips are the shoots or sprouts that grow from a sweet potato tuber; planting slips is the standard way to grow a crop of sweet potato tubers. Uncle Clif ships half a million of them around the US every year.

It's funny how so many of us farmers who spent our childhoods around or in farms swore we wouldn't grow up to farm. Uncle Clif's father was a disciplinarian who put him and his siblings to work on their farm as kids, telling them they couldn't go out with their friends because it was a "key to the jailhouse door." Uncle Clif remembers that when he was sixteen, his friends were going to the beach for Memorial Day, but his father wanted help with the sweet potato crop. "He said we couldn't go no matter what. I told him, 'I can tell you one damn thing. When I get grown there's two things I ain't gonna do. I'm not gonna farm, and if I farm I ain't gonna raise no damn sweet potatoes.'"

Uncle Clif's high-school principal pushed him to be accountable, too, catching him with a bottle of wine right before graduation and threatening to hold him back unless he enrolled in college. Uncle Clif went to Virginia State University, but not with the best of intentions. "I majored in agriculture because my thinking was, I won't have to study, I'm going to have me a good time partying." The principal was onto him. "He said you can go up there and act the fool if you want, but the Slade family name will be a shame and a shambles."

Uncle Clif's roommate was my community neighbor Uncle Roland Terrell, who also had no intention of staying in school more than one semester. When December rolled around they changed their minds. "We argue to this day who said it first: 'I'll come back if you come back.' I'm glad somebody said it. The rest is history—we finished, both of us got our degrees. My daddy was never any prouder. He had a seventh-grade education."

That was 1975. Uncle Clif went into teaching and extension agent work and became respected for the way he helped other farmers through some bad drought years in the late '70s. "I have sat down with Black farmers and white farmers at the end of those disaster years and fabricated a plan to let them farm another year. I noticed that the white farmers, they always wanted me to come to their house, but they would always pick a time near dark. They didn't want me to be seen leaving their house."

Farming was becoming much harder to sustain financially, and he saw that farmers his dad's age, who often weren't educated, were having a hard time keeping up with the management skills that were beginning to be required. "I remember when my daddy didn't owe anybody anything," he says, "but an agent sold him a diesel tractor, and he was $100,000 in debt. Then we started getting these droughts." Uncle Clif saw that banks were making bad loans to farmers. "We had a lot of Black-owned land back in the '60s and '70s. A lot of farmers mortgaged that land and the lending institutions loaned them money in hopes they couldn't pay it back, and a lot of them didn't." In his county today, the two biggest farmers are white. "They have seven to eight thousand acres each, and it was land that used to be owned by minorities."

Despite all that, the Slades did hang on to their land, and Uncle Clif came back to farming himself. "It was about 1980 before the magnet pulled me back," he says. He asked his father if he could use a small part of his land. "He said 'Sure Sonny buck, but whatcha you gonna use it for?' I said 'I'm not sure but I think I'm going to raise me some sweet potatoes.' He took the back of his hand and put it to my forehead. He said 'You okay? You all right?'" Uncle Clif's father couldn't believe he would choose to farm sweet potatoes.

Even though Uncle Clif's father had had a rough time with farming, a few of his crops were moneymakers. He made his mark in seed catalogs, too, which mentioned that Uncle Clif's father grew a regional heirloom variety of sweet potato called Virginia Baker. "The week of Thanksgiving, my dad would sell three to four thousand dollars' worth of potatoes and collards right at his house. The money was rolling. That was his niche of having money for Christmas. He was an innovator with a seventh-grade education."

Throughout his career Uncle Clif has looked for ways to maximize revenue on his farming operation. When he worked at Virginia State University, Dr. Jewel Bronaugh was one of his colleagues, and she once challenged him to think outside the box to make his farm work harder financially. "It kind of upset me the way she provoked me, and I thought about it all the way home on my hour and fifteen minute commute. How can I make more money doing this stuff? Let's break it down into square footage." He came up with a goal of earning $1 for each of the 43,560 square feet in one acre of farmland. He methodically eliminated crops that weren't space-efficient and weren't profitable, and that helped him focus on a few key crops. "I ended up with sweet potatoes, sweet potato plants, and elephant garlic."

His stick-to-it-ness, innovation, and agricultural tenacity have resulted in success. After developing his full-fledged business producing sweet potato slips, he could generate five or six figures of revenue in two to four months' time, between March and June. For about a decade, he became the go-to farmer for slips, reading the market and pursuing organic certification, maximizing every aspect of community interest for the growth of his third-generation farm.

In some cases Uncle Clif has been able to exceed his goal of $1 per square foot. "Last year I sold $60,000 in sweet potatoes just from one acre, and nobody taught me that from any school. I'm still studying ways to make my garlic bigger and better. My goal is to sell $150,000 of garlic in an acre and I'm two years away. I'll get there." He has grown slips and seeds for Southern Exposure Seed Exchange for over a decade and is presently growing some African plant seeds for a Black-owned cooperative company Ujamaa Seeds. "As I started selling the plants, that's when I found out Africa's contribution to sweet potato production, and that made me prouder to do what I was doing," he says. "All the varieties of seeds I'm growing for Ujamaa originated in Africa, and I want to be part of it."

After over forty years as a commercial grower, Uncle Clif is still learning and innovating, figuring out ways to add more revenue to his operation. He and I have talked about the untapped potential of selling the leaves of sweet potato plants. "You can talk about doubling your money," he says. "There's $20,000 per acre in the leaves with the right market, and eating them can help control blood pressure. And at Virginia Tech, they're pelletizing the stems and feeding it to tilapia."

Uncle Clif also wants to nurture the next generation of farmers. Recently he agreed to let a younger farmer work with part of his land and sell her products under his name. It's good for them both: She'll help him with social media and he'll mentor her on the growing and the business side.

"I think she could take sweet potato plant production and sales to a new level on this farm, so that's the opportunity that we present to each other," he says. "When I gave her the opportunity she started crying. She said, 'You joking.' I said, 'I'm not joking. Somebody had to start me off, you know?'"

I'm obligated to do the same. I don't exist as an agriculturist without my older trees. They kept agriculture alive for those of us in my generation. Specialty crop farming is a little more chic these days in the age of social media, so we don't have to struggle with all the same barriers, though there are still plenty of challenges that haven't changed over the decades. Sawubona Mr. Clifton S. Slade Sr., Mrs. Ernestine Wells Slade, and Mr. Clifton A. Slade.

MR. DURON CHAVIS

Mr. Duron Chavis is a a great friend of mine and fellow farmer in Richmond, Virginia, who grows food, a lot of food; but like me, he's also growing a new crop of younger farmers. We connected immediately as supportive comrades when I met him in 2018. He's gone from starting a few urban gardens to founding a community land trust and developing Happily Natural Day, a community-based and vocational nonprofit that organizes an annual festival called Happily Natural Day (now twenty years running) and a training program for new and beginning farmers. He probably doesn't know that he was the initiator or sales representative of my first sale for Carter Farms. He orchestrated the sale of tropical seedlings I was growing in my parents' basement to the organization he was working with at the time. It's a mycelial relationship—we're constantly and consistently feeding off each other and allowing that to feed others, and both of us were fed by some of the same elders, including Baba Azibo Turner in Virginia and Baba Oduno in DC. Sawubona Baba Oduno, Baba Azibo, and Sister Latifah.

Baba Oduno was a member of the Washington, DC, chapter of the United Negro Improvement Association (UNIA), whose meetings I attended, and he was a big help and inspiration to Duron Chavis, who was known as Brother Manifest when he founded the Happily Natural Festival in Richmond in 2003—an event that celebrates natural hair and the diversity of Blackness. Through the UNIA, Duron had traveled up and down the East Coast and met a network of elders, plus young people, who were practicing agriculture as a form of activism.

"We were doing this work of building Black liberation and empowerment, building Black owner-controlled spaces and systems in our communities," he said. Duron had not grown up farming, but he met Black farmers—including Mr. Renard "Azibo" Turner—when they came to Happily Natural.

"I'm a city boy," Duron says, "but I fell into this farming thing backward. Farmers kept showing up and playing the role of mentors, and I got deeper understandings of why agriculture was so important." Brother Azibo gave him a sense of urgency about the work, Brother Saleem Ahmed taught him how to make farming work as a business, and Baba Oduno helped him see how agriculture was connected to a larger movement. "Those three Black men were like 'Yo, young man, I see you out here; let me help you evolve in this work.' It's like they were giving me pieces of a puzzle."

Mr. Duron Chavis began to help with distributing produce in urban communities, and in 2012 he started growing food himself, on a vacant lot acquired by the city of Richmond. Most sites in the city of Richmond and most cities around the world have heavy metals in the soil, plus excessive amounts of concrete, cement, other debris and trash that keeps that soil from being productive. Brother Manifest couldn't plant directly in the soil, so he created raised beds that he filled with compost and potting soil.

"It took off and evolved," he says. "We realized we needed more space, so we found larger plots with landowners, kept evolving in our practice and building our skills of how to grow food." He established Sankofa Urban Orchard, five acres of green space and fruit trees in the city, plus other urban gardens and youth gardens. He went on to gain access to other larger properties, with dozens of acres, outside Richmond.

Duron took advantage of grants and founded a nonprofit to build up this infrastructure—not only the land but cold storage, refrigerated trucks, and equipment. He is interested in profit-making businesses that are sustainable long-term. Baba Oduno told Duron, "You got a community of people, so build your ecosystem." For nine years, he's run a training program to nurture the next generation of young farmers—thirty or so a year—and given them the chance to grow their new businesses at incubator farms. "How do I plant these seeds?" he asks, "How can I be the Oduno or Ahmed for the young people coming underneath me?"

It's about the people more than the plants. "We create a space where folks are able to talk and learn from each other," he says. "I thought we was going to be training farmers, but half of these folks ain't gonna farm. They might be people who do events. There's a whole clique of people who have been in my trainings, and they're bringing out the yogis and DJs and music and seed starting. They're having a blast, supporting the farmers that are actively farming and bringing more volunteers. All of these people are arranging themselves."

At least one of Duron's students is now successfully running a farm and nonprofit in Petersburg, Virginia, on land that Duron acquired and then leased him. This young man, Mr. Tyrone Cherry, is now training other young people to farm in an agricultural training program he calls Young Pharaohs at the Petersburg Oasis Youth Farm. "For me, that's success," Duron says.

Brother Duron is a force in the urban ag landscape, and he's also a community conductor, composer, and arranger. He's always been a bridge that connects people, land, food, vocations, the past, and the future like a dynamic music producer, sampling classic beats and adding them to modern rhythms. And he's working on issues such as affordable housing, urban heat islands, and food deserts, too. At an eighty-acre farm plot in Amelia County, he envisions developing housing for farmworkers and just for folks who need housing and might enjoy the farm as an amenity. His goal is to make food production culturally relevant for Black people through music, art, and events. His tree is producing so many different fruits, dropping all over the region and the country.

"For me, this is holistic work," he says. "None of these things are disparate; they're all interconnected. The garden is art by itself, but we actually put art in the space to attract people riding by. When they see the murals of these revolutionaries, they ask, 'Who's this?' Fannie Lou Hamer? Shakur? Carver? Putting that art up is a tool for political education. Once you come you see the fact that in the middle of Southside Richmond, there's a seven-acre farm growing food in abundance, and there are Black people doing this. It's not some white nonprofit. We're doing it in a way that's culturally relevant." Sawubona Mr. Duron Chavis and Brother Manifest. Sawubona Mr. Tyrone Cherry, Mrs. Natasha Crawford, and Brother Obar.

B. J. DENNIS

The Lowcountry in South Carolina is a special place for African American history and foodways. It's home to the Gullah Geechee people, who have maintained a unique culture due to their isolation on the Sea Islands off the coast. As described in chapter 5, rice played a big role in the history of that area, and it's the foundation of Gullah cooking as well. Chef B. J. Dennis, known these days as the Gullah Chef, has deep family ties to the Lowcountry, and he's been spreading the word about rice and Gullah cooking.

"Growing up," he says, "you didn't realize what you were looking at up and down the coast with old rice fields and dikes that's amongst us throughout the Lowcountry." His grandfather cultivated an acre and a half of rice behind his okra field, but the old rice fields, created and tended with tremendous amounts of Black labor and knowledge, are not so visible these days.

"You would think in South Carolina you would be taught about rice culture," Mr. Dennis says. "You look at these 'planters' and these enslavers who have monuments and streets named after them. The truth is that the enslaved were the planters who brought the knowledge here. When I was nineteen, I started reading books to unravel and see the truths. I would say that the disconnection is done on purpose, truthfully."

Mr. Dennis helped me to further understand the spiritual aspect of rice for West Africans. On one of his trips to Senegal, visiting with Jola and Balanta people, he noticed that the rice fields looked just like the fields at home. The name Balanta in Mandinka means "those who resist," and that spirit is reflected in the resistant and ever-present rice culture, traditions, and history of the South Carolina and Georgia coast. "We brought those same systems from there to here," he says, "so it's reconnecting through that lens, the whole spiritualism behind it, even though it was weaponized against us. They wouldn't sell me rice in Casamance, which is one of the older rice-growing regions in the world. They *gave* it to me because it was spiritual to them. They were dancing with rice in their hair—it was a celebration."

I've historically witnessed that in every aspect of the African American experience, things that bear spiritual significance—like the drum, cotton, or rice—are weaponized against us. Mr. Dennis told me about tasting thieboudienne, the national dish of Senegal, while he was there. This dish is the Goddess Mother of jollof rice. Benachin is the Gambian version of thieboudienne or jollof rice. "Benachin is next level," he says. "They use white bisap"—which comes from hibiscus—"and white sorrel in the broth. In Casamance that's where you see the OG [original] version of benachin, not the red version but the brown version." These dishes that develop over time with local ingredients are a cultural treasure, and African rice dishes have been adapted in the Americas to other signature foods, including Hoppin' John and chicken perloo.

"They are still growing ancient varieties of rice over there," Mr. Dennis says, "black rice, pink rice, red rice." He'd like to revive an awareness of historical rice varieties—both in Africa and in the United States—and to lessen the stigma around rice. "My work has really moved from more of a chef to culture-bearer work through food. It's about repatriation of our cultural food, because there's a disconnection from agriculture; it's a trauma thing for a lot of us. Our foodways have gone to the wayside because of separation from our agrarian roots. At one point in Charleston they documented 100 varieties of rice. The heirloom heritage rice from West Africa, our ancestors grew in secret gardens." Super white, bleached rice from grocery stores is a poor replacement for these diverse rices.

Mr. Dennis started off as a curious child with an interest in cooking and worked his way up through restaurant kitchens from Charleston to the Caribbean. Eventually he started hosting Gullah pop-up meals where he'd sometimes cook over an open fire or pit roast a hog, some of them on land owned by Gullah organic vegetable farmer Joseph Fields. He's incorporated more research into his cooking—studying ingredients, recipes, techniques. Gullah cooks have created many of the most beloved Southern dishes, and those foods go directly back to Africa.

"The cultural heritage work is the unpopular work," he says. "It's not what's trendy, it's not what makes people comfortable. Sometimes you gotta read books you don't want to read, but they have jewels in there if you filter through the fluff and BS." As an example, Mr. Dennis mentions *A Woman Rice Planter* by Elizabeth Allston Pringle, who grew up in the Civil War and wrote about how she wanted to grow rice after the war but couldn't do it without the "darkies." Mr. Dennis says, "But that book also has stories from royal families in Africa."

His goal is not to become famous. It's to make sure the Gullah contribution to American cooking is recognized properly. "Authentic work is done till you die," he says. "I wasn't going to allow them to make me trendy." When Mr. Dennis was approached about writing a book, he humbly suggested that Emily Meggett, a Gullah cook from Edisto Island, be the first to publish a Gullah cookbook instead; her book, *Gullah Geechee Home Cooking*, came out in 2022.

Like myself, Mr. Dennis and Brother Manifest (as well as the others noted in this chapter) are active readers and researchers, curious about stories and accounts from and about our ancestors. Research, both on the farm and through books, cookbooks, census records, deeds, and elders' accounts, drives our work and our passions. Sawubona Mr. B. J. Dennis. Sawubona Ms. Emily Meggett. Sawubona Jolas. Sawubona Balantas.

JULIUS TILLERY

"I've farmed cotton all my life with my dad," says Mr. Julius Tillery. "Before it was a farming business for me, it was just the chores. I've been involved with the farm since I was five years old, but it wasn't Black Cotton until 2016."

Black Cotton, the multigeneration cotton farm Brother Julius runs in Northampton County, North Carolina, is changing the perception of cotton with African Americans one boll of cotton at a time. Just like my family, his family has owned their land since the early twentieth century, and historically they've raised soybeans, corn, and peanuts. After Julius's grandfather died in 2017, he's become a bigger part of the operation, working with his father.

"I think the intention was for me to do other things, something seen as successful," he says. "When you're a smart kid growing up in this area, the absolute last thing people say is 'You should be a farmer.' I think my intentions changed when I realized how important I was to our family farm enterprise. I worked other jobs for my enjoyment, but I would take 80 percent of my vacation days to work on my farm. Farming's not vacation for me, it's responsibility. Those taxes had to be paid."

Brother Julius's devotion to his farm, his legacy, and the cotton is a testament to his upbringing and vision.

He and his father (who, like my father, was an agriculture teacher) raise fifty acres of cotton and seventy-five acres of soybeans. "I remember saying to myself, I have to make the farm more valuable," he says. "I tried shiitakes and vegetables and still grow those, but it's very labor intensive." Brother Julius began to think about how he could transform the farm, which led him to wonder about growing cotton.

Just as I'm trying to lessen the stigma around farming for Black people in general, Julius is working hard on reducing the stigma associated with cotton. "Cotton is one of the crops that make people say, 'I would never farm,'" he says. "It's those negative slavery roots, tropes about an oppressed life and a poverty-stricken life. People associate sharecropping with cotton, too. People who were still farming in the '70s and '80s, they were not proud of being farmers."

Julius has found ways to grow his family's farm while being visible and proud as a Black cotton farmer. Social media is a big part of that. A few years ago, he recalls, "It was a bad year; it was the worst cotton crop everywhere. But I was proud we had something, and I videoed myself, saying, 'This is my Black farmer challenge. If you got some cotton show it to me; I bet you we can't find five Black farmers got a better crop than us.'" Julius's father warned him not to show the crop, but he put up the video online anyway, and nobody else came forward to compare their crop to his.

He began to realize how rare it was to be a Black farmer who raises cotton, not only because of the stigma but because of the high operating costs. Speaking at conferences, organizing events, and doing social media helped him build a brand while figuring out nontraditional ways to make money from cotton, such as selling wreaths and home decor made from cotton bolls. This led Julius to living one of my college-era dreams, growing his own cotton and making his own shirts. The Vans company took notice and approached him about a collaboration on T-shirts made from his cotton and featuring his logo. The shirts dropped in 2022.

It is powerful for a Black farmer to be owning the image of cotton like that. "We have a place in this industry," Julius says. "We are the tastemakers of the country, the leaders in fashion and what's cool and what's not, and I think we should be able to see the value in us producing these things. The work of our ancestors was so important in creating this value that we should show appreciation."

Another thing Julius and I have in common is that some of the sources that could have taught us pride and entrepreneurship in farming—such as our agriculture teachers and the universities we attended—didn't present farming as a viable occupation. And we're both trying to change that.

"When I first started," he says, "People would come and see the farm, and my father would say, 'Why do you bring these people here? There's nothing to see.'" But Julius can see that touring the farm engages the imagination of visitors and radically changes their understanding of working with the land and what a life as a cotton farmer can provide.

Mr. Tillery says, "I don't have enough capabilities and resources to make up a big lie about cotton farmers; I don't have nice equipment, I don't even have a barn. I'm a poor farmer, but people perceive me as quite wealthy and rich because of my influence and spirit."

Like me, he's had to grow out of some of his own perceptions about farming. "I know that I'm going to be a lifelong farmer. If I'm going to do this until my last year, and I'm thirty-nine now, how can I make this work for us so we don't look at it negatively all our lives? If I don't, the next generation's going to completely give up." Sawubona Mr. James Tillery Sr., Mr. James Tillery Jr., and Julius Tillery Sr.

P. J. HAYNIE

The first African American to purchase land in Northumberland County in Virginia was Mr. P. J. Haynie's great-great-grandfather, Mr. Robert Haynie. He didn't waste any time after emancipation—this was 1867. "It inspired me," P. J. says. "I wanted to continue my family legacy of farming, and I wanted the opportunity to get my children what I experienced growing up—the independence and freedom to do the Lord's work."

He grew up in the heart of his five-generation family operation, raising corn, soybeans, and barley. "I tell people, my daddy tricked me into farming, because at ten years old, most kids were getting toys, but my daddy bought me the real thing—a real tractor. Wherever Daddy's crew was working, they'd have my tractor parked in the field hooked to the same implements the forty- and fifty-year-old guys were using. If they were planting, I was planting; if they were plowing, I was plowing."

It put him in a position to be confident at an early age. "In high school I got involved in FFA and participated in tractor driving as a freshman. I beat the other kids. To have them calling me nicknames, and the jokes and slurs we dealt with, I wanted to prove we deserved our place at the table. Daddy always told me you might not be the richest kid, you might not be the smartest kid, but you will be the hardest working. Nobody's gonna outwork you."

At Virginia Tech, P. J. studied agricultural economics and took on leadership positions—something he's continued as an adult, serving as chairman of the National Black Growers Council and a prominent advocate for Black farmers in government and private industry related to large scale agriculture and Black farmers. His education helped him analyze the business side of the family operation. "I figured Daddy knew how to grow it and I can help him sell it. Growing up, Daddy always told me, a bad business is better than a good job, and I'd say that's the craziest thing I ever heard. But it's true, 'cause ain't nobody gonna walk in and fire your dumb ass. You set your own destiny."

As soon as Brother P. J. shared this inspiring, intelligent quote with me—"A bad business is better than a good job"—I immediately told my sons about it. It's how many of us have lived, and the quote verifies how in tune we are with entrepreneurial spirits.

He has embodied that ethic in farming several thousand acres and running multiple businesses (including bulk transportation, timber harvesting, and landscaping companies). And in 2021, he made the leap to purchase the Arkansas River Rice Mill in Pine Bluff, Arkansas. It's now one of the few Black-owned rice mills in the country, capable of milling a tractor trailer's worth of rice every hour.

"I've been able to move my family from the feed business to the food business," he says. "Forty years ago, Granddaddy sold corn to Tyson and Purdue in Virginia for $4 a bushel. Twenty years ago, Daddy was selling it for $4 a bushel. Presently I'm selling corn for $4 a bushel. The problem is that Granddaddy's combine cost $60,000, Daddy's cost $150,000, and mine cost $700,000. I gotta figure how to get more money for that corn. I see downstream opportunities in food that I didn't pay attention to when I was in feed."

P. J. knows that the demographics of farming are tough and that the barrier to entry is very high. He feels that the greatest gift he can give to his family is for his children to enjoy the same lifestyle as he has while farming, if that's the path they choose.

P. J. is well aware of the inequities in American land ownership and how that affects generational wealth. "I participated in a documentary film that Al Roker produced called *Gaining Ground: The Fight for Black Land*. It documented how over the last 100 years African Americans have lost 14 million acres of land," he says. Figuring on land value and the value of agricultural production on all those acres, "it's over $400 billion gone and removed from the Black agricultural community that we will never get back. So that's the dilemma."

Much like Ms. Ebonie Alexander of the Black Family Land Trust, Brother P. J. believes in making sure younger generations know this history. "I think the future is to make sure your children and grandchildren understand the value of land—owning a piece of that rock. I don't care if it's just a garden in the backyard where they go after school, raise vegetables, and put out a roadside stand. There's no greater joy than watching a seed you plant grow, and harvesting the mature crop, something you nurture." He also stresses how important it is to legally structure land ownership so that it doesn't slip away.

"Grandmama and Granddaddy did not want to show favoritism to those children, so they left that farm to all of them." They didn't put the land in trust or an LLC or a corporation, P. J. explains, and now the land is left to more than thirty grandchildren, who will have to sort out the future of the land.

His advice to younger farmers: "Find the network of people who look like you. They may not be close, they may be a thousand miles away, but they're people you can share ideas and techniques [with] because you are going to have to do things differently."

Brother P. J.'s inspiring family history speaks to his deep roots that have created the abundant fruit. It all goes back to the Mr. Robert Haynie tree, first germinated in 1823 and transplanted to his own land, purchased in 1867. The richness of their story, their persistence and devotion to the craft, and their cultivation of their legacy motivates and encourages me. Sawubona Mr. Robert Haynie, Mr. P. J. Haynie Jr., Mr. P. J. Haynie III, and the entire Haynie family.

BONNETTA ADEEB

A few years ago I rode to a BUGs (Black Urban Gardeners) conference in North Carolina with Mama Ira Wallace of Southern Exposure Seed Exchange. On our four-hour car ride, Mama Ira shared with me how, after decades of working in the seed business through Southern Exposure, she was hoping that other Black growers, gardeners, and farmers would take up her work of seed preservation.

Mama Ira's words stuck with me. I didn't want an elder's words to fall to the ground, especially one who has had such a strong impact on maintaining legacy via seeds. This conversation influenced me to start a seed business, Carter Brothers, with my sons, to sell African seeds. Not long after that, I met an elder sister at the Taking Nature Black conference in 2020. What started as a standard courteous "How are you" has grown into a thriving collaboration. Mrs. Bonnetta Adeeb has been going and growing ever since. At the time, she was running a youth services nonprofit called STEAM ONWARD. A few weeks later she established an initiative for seed collection. The Ujamaa Cooperative Farming Alliance, which began as a sub-program of STEAM to focus on BIPOC seed growers, heirloom seeds, and culturally significant crops, is named after the fourth principal of Kwanzaa, the Swahili word *ujamaa*, which means cooperative economics.

The evolution of their movement happened fast, and Auntie Bonnetta has a network of inspired, motivated individuals, young and not so young people. The organization moves like an army tackling a mission. She has a solid team and approach and is able to network at every level of the industry as well as bring in research institutions along with the private sector.

"I was born into the Marcus Garvey movement, as was my husband," Mama Bonnetta says. "It was very clear that the wisdom came through the elders and ancestors. I remember being three years old, and my aunt showed me a book of what they discovered in Egyptian tombs, and saying, 'Your ancestors built the pyramids.' We understood that if you are standing tall it's because you stand on the shoulders of ancestors."

She and her husband, Mr. Hassan Adeeb, moved to DC to research and write a series of books on African kingdoms and to work with high school children. Two Howard University professors became their mentors, encouraging them to view themselves as scholars, not just as teachers. "They told us, 'We have kids coming here to major in African studies and they can't name two African countries. We need y'all to be doing the work at the precollegiate level so they are not so doggone ignorant by the time they get here.'" Through Howard, they both traveled on Fulbright scholarships to Africa and continued learning the stories of their ancestors, and feeling spurred on by ancestral voices. "They won't leave us alone; they make us do it," she says. "When we're on the right track the blessings and rewards that the ancestors and the Lord send our way is immeasurable."

Today, she focuses on growing food. "It's a new renaissance" for Black farming and gardening, she says. Now, her work is to build a collective of BIPOC seed growers and to give young Black children crucial early exposure to growing food and plant breeding. Ujamaa runs the Ira Wallace Seed School and other workshops, maintains nine regional seed hubs around the country, and runs a four-acre seed farming plot. They also sell seeds through their website Ujamaa Seeds (ujamaaseeds.com).

"I call collard a gateway crop or a gateway drug," she says. "Even if people have never tasted it they think they love it, and they will grow collard greens. We've got an heirloom collard project we've been working on for years; we have 100 varieties and we're breeding new varieties." Ujamaa also offers a curriculum called Collardz 4 Kidz, which guides children as young as age three through the steps of growing collards and saving the seeds. "It teaches the science of plant breeding, soil science, and plant biology, and all the kids learn how to start a business and make money off collards they raise." Ujamaa has produced three books on collards as well. The books are packaged with seeds, a grow bag, and instructions. "So, when Mama reads the book to the children, Mama learns the science of that breeding."

Auntie Bonnetta and I have discussed building US markets for African vegetables. She says, "Many of these foods can reduce some of the diseases we have because we don't have access to our traditional food." (Sound familiar?) She's trying to inspire people to get to know these crops—some of which they recognize only as ornamental flowers. "Now that we have people excited about collard greens, we're sneaking in what I call 'the Africans,'" she says. "We've been publishing literature called the African cousins, building on the concept of three sisters. Do they want to eat cleome or celosia? We're asking people to grow them in the same garden where they would grow collards." Along with that come egusi and other African ingredients that are often added to greens and stews for flavor.

Ujamaa is founded on spreading skills and knowledge that's rooted in heritage. I asked Auntie Bonnetta what the significance of okra and sorghum are to her and she answered, "Ownership. Having something that belongs to you. Black people think of themselves as orphans and an orphan is a sad thing. Many orphans believe their parents gave them up because they didn't love them. So, what okra and sorghum mean to me is the proof that Mama did love her babies. We believe that Mama braided those seeds in the hems of the garments or in the hair, because she had hope for them. Sorghum is a grain, and it's very much like us—we don't die, we multiply. Sorghum started out in North Africa and it naturalized itself all over the world. If you think about the African diaspora, Africans are now in 117 countries outside Africa, and wherever we go we have an influence on the culture, traditions, and everything else."

Auntie Bonnetta identifies as an educator of children, so she's very intent on connecting with younger generations. "So much of what I believe this generation is about is linguistics," she says. "If you look in our catalog, most of the peas are called African peas instead of cowpeas or field peas. . . . We're reclaiming the heritage by changing the language around it. We like the phrase *self-determination*—we have the right to name ourselves. You can call it cowpea if you want, but I'm gonna correct you, it is an African pea."

In 2022, a Tuskegee extension staffer in Maryland asked her if the Ujamaa farm could use some help from young men living in a Northern Virginia detention center. "The boys have been coming out on Saturdays for three years now," she says. "What a blessing they are to me. They love the farm, being out in fresh air. They mow, they plant; they planted almost 500 trees last year. We are hoping to turn that into a program to give them some type of certification. They say they want to come out to us because we like them and we give them snacks and know them by name and care about how they're doing."

She thinks these young adults could develop their entrepreneurial spirit—also part of their heritage. "When you're on the streets of any city in Africa, everybody is entrepreneurial."

Growing food is one of the best ways Black people can connect to their heritage, Mama Bonnetta says. "I had somebody tell me they couldn't grow nothing, they had a black thumb. Why you tell Bonnetta that? It's like, of course you do, girl, it's in your blood to grow your own food, your folks built the economy of the United States. Of course you have a Black thumb—and it can help you survive." Sawubona Mama Ira Wallace. Sawubona Mrs. Bonnetta Adeeb and Mr. Hassan Adeeb.

DR. GLADYS WEST

I'd be remiss if I didn't acknowledge the genius, humility, and brilliance of mathematician Dr. Gladys West, contributing developer of the Global Positioning System (GPS) while working at Dahlgren naval base in King George, Virginia. Dr. West was a friend and fellow graduate of one of my great cousins, Aunt Reba Galloway, at Virginia State University. I had the wonderful opportunity to visit Dr. West and her husband, Dr. Ira West, at their welcoming home a few seasons back. Farmers rely on GPS guidance in their tractors for accuracy, and every USDA office in America uses GPS technology to map out where farms are located to determine acreage accounting, and GPS is at the core of precision agriculture. GPS has ushered in a new wave of technology, convenience, and opportunities. Sawubona Dr. Gladys West and Dr. Ira West.

MEKHAEL, YAHIR, AND SHMAEL CARTER

It's one of my greatest goals to make the world a more welcoming place for my three sons, Mekhael, Yahir, and Shmael, as farmers and agricultural entrepreneurs if they choose to do that work. They're my literal fruit, and I've tried to make sure they have a strong connection to our family story, to our history as farmers, and the land itself.

From 2019 to 2023 they ran the Carter Brothers seed company, selling amaranth, kale, corn, managu, Nigerian spinach, and other types of seeds. They raised some of the seed themselves, and they also sold scented hand sanitizers and other value-added products. They did well in their sales endeavors. "We had our own logo," Mekhael remembers. "We made pretty good money. It was good experience." Mekhael handled the marketing for the business and was the cashier at events where they sold their seeds and products. "I was the marketing person, and I like doing the cashier roles," he says.

These days, they're helping me out on the farm—planting, weeding, fixing fences, and helping with our cow. Mekhael tells me it's fun, but he acknowledges that he likes it better when there's a light rain to take the edge off the heat.

I've tried to teach them some skills and also to show them that there's a future in agriculture. I want them to think of farming as a family legacy and also as a real career path they could be proud of. I think I might be succeeding. Yahir is dreaming of a career in the NBA or FIFA, and he says he thinks of farming as an option that's there for him.

At sixteen, Mekhael already has a sense of those who came before. "When I'm working, I'm just thinking about the task, but it's nice when your grandparents bought land and you're working on it now. I'm appreciative. I think I'm going to end up in [farming] most likely. Even if I'm not doing it full time, I'll do it part time. It makes good money. I like the learning part, being outside." Mekhael says he is also interested in working in a food-related industry, such as running a grain mill.

"You will know the tree by the fruit it grows" is a common proverb in many cultures. These are just a few of the fruits of our beloved industry and ancestral calling—family, responsibility, and industry. Whether a fruit takes the form of a legacy farmer, a first-generation farmer, an activist, a historian, a scholar, a researcher, or a passionate consumer, that fruit represents a legacy for the future.

Just as I can describe only a few fruits here, the field of study I call Africulture has only scratched the surface of the principles, practices, plants, and people that have contributed to agriculture, as well as the policies and psychologies that have stood as challenges and obstacles to melanated farmers. Africulture continues to grow and evolve daily, as historians, farmers, chefs, agronomists, cliometricians, researchers, content creators, and many others share and reveal how the seeds in the fruit of our ancestors are evident in the culinary, cultural, dietary, and historical traditions that we still honor today. Social media provides historical context and knowledgeable agricultural content in three minutes or less, and has played a tremendous role in making the next crop of future farmers more aware of the practices, traditions, and foodways of our ancestors.

The Bahamian minister Myles Munroe preached that anger helps in revealing your purpose. I, along with many of the people I mention in this book, have felt frustration bubble out into anger related to the vocation of agriculture. Our anger and frustration has fueled us and we have used it to refine our focus on working for and with our fellow farmers who haven't always seen a fair growing field. Our purposes have flourished through a controlled, intellectually and emotionally managed episode or miniseries of anger. In a purposeful sermon on the book of Proverbs, chapter 19, Minister Munroe expounds that there are many plans in a man's or woman's heart, but it's our Creators' purpose for us that will ultimately be successful. Many have chosen agriculture as a profession, but for myself and a cadre of others, our Creators, our ancestors, and the land have chosen us.

If all hearts and minds are settled . . .

I want to thank (quite sarcastically) all the forces, entities, people, and policies that have tried to suppress and disenfranchise Black and Indigenous farmers, land stewards, and pastoralists. And I'd like to sincerely thank all the Black and Indigenous farmers who have survived in the face of those obstacles and maintained the legacy, dignity, and value of the farming occupation.

Africulture will continue to grow. Its roots will stretch out longer and deeper. More seeds will germinate in places one never would have expected. It is a force that will overcome pestilence and obstacles like a great tree would do, and its fruits will be full of sweet, nutritious, life-enhancing seeds: a legacy for generations of future forests to grow from. Let's grow!

The law of use states if you don't use it, you lose it. I'd like the United States to stop using racism and discrimination so much, so eventually we can lose it. I am deeply frustrated to see racism being conserved and recycled in so many ways. It seems as if every generation is handed a rejustified, repackaged version of the same contaminants. Growth generally requires losing or shedding that which inhibits growth, and people of African descent need to finally have the opportunity to grow without restraint.

Historically, Black people have thrived and competed well in all forms of races. Reverend Jesse Jackson notes that "the reason why black and white can play together is because the playing field is even, and the rules are public, the goals are clear. . . . Beyond the ball field, the playing field is not even, and that's our challenge. That's all that we want in America, is an even playing field for every American."[1]

The systemic psychological, social, and financial challenges that African Americans and people of African descent have faced has been and continues to be documented via research publications, articles, and books. My goal in Afri-culture—both the organization and this book—is to highlight some of those historical challenges that may not be widely known, especially related to the field of agriculture. As award-winning author Ms. Octavia Butler writes, "All that you touch you change. All that you change changes you." It's up to those who have historically perpetuated and spread this disease of discrimination and prejudice to commit to change if they want it to occur. My community cannot do it by ourselves. We have tried, followed all the rules, adjusted to all the shifting environments, scored goals even as the goals shifted, yet the rules still change and shift. As we move from African intelligence and agricultural intelligence to artificial intelligence, I'm not sure how much time we have left to get this right. We've proven we can endure, we've proven we can adapt, we've proven we can grow through whatever to produce and provide flowers and fruit. Now, I call on this nation I call my home to deliver justice to its most deserving and enduring citizenry. It's been my fervent desire that positive change would happen swiftly, and at moments, that seemed to be unfolding. But now, looking back at history and comparing it to current events, I feel that positive change is unlikely. What is likely is that Black people will continue to contribute to this country agriculturally, adding flavor, culture, and massive influence, with or without recognition. What you are not changing you are choosing. America has willingly chosen this reality for its long-term residents and citizens of African descent. To deliver different results, different realities must be chosen. America's true greatness is directly tied to how its citizenry chooses to treat people who look like me. Can America really be great, or will it accept the mediocrity of this social structure that selectively provides favored citizen status based on color-coded parameters?

As Mr. James Baldwin famously said in the 1960s: "To be a Negro in this country and to be relatively conscious is to be in a state of rage almost, almost all of the time." I'd like to live a moment of my life in this country *not* in a state of rage, while maintaining my level of consciousness and awareness. At moments I feel like the fictional character Eddie Kane, from the movie *The Five Heartbeats*, who sings at the movie's end, "I feel like going on." Many other times, like Marvin Gaye, I just wanna holla. Music is a part of my therapy; it helps me remember that I'm not the only one hurting, but it also lets me know that my people have been dealing with these things longer than we should. Much of the music I listen to and quote in this book is older than I am. It's considered timeless music, and it's timeless because many of the conditions these artists sang about haven't changed. It's a lyrical time capsule of anger, frustration, and a plea to the humanity of what appears to be a country without ears or a soul.

Growing up in the church, we were instructed not to let anyone steal our joy, and that joy would come in the morning. But what if you were born with no joy, or suffer from the American-induced joy deficit disorder? I so wish there was no need for me to write this book; I wish that the previous season of books about the Black experience in America would have been enough. But it wasn't, it may never be enough in my lifetime, and that's concerning to me.

There is a sincere appreciation of sharing this information with students, readers, and audiences across the country and around the world. There is also a frustration with why certain basic aspects weren't made part of the educational curriculum and cultural upbringing of this country long ago.

Curtis Mayfield and the O'Jays sang to people to get ready, there is a train coming, and that the train that was coming was a love train. A train containing the love of the land, the love of our environment, the love of cultures, and the love of humanity, despite their physical complexions.

Africa is the cradle of agriculture as well as civilization. As you finish reading this book, I invite you not to stop learning about the love train. Keep researching and adding to Africulture. There are so many more stories of the principles, practices, people, and plant relationships that arose from Africa and continue to contribute to agriculture in the United States. I've just barely scratched the surface.

May these words, experiences, examples, metaphors, heroes, and sheroes be an inspiration for you and your decision-making processes that inspire the next generation of agricultural influencers and change makers that the world needs to experience.

ACKNOWLEDGMENTS

itYAHfaot
i thank YAH (Hebrew for the Creator) for all of this.

This is a phrase with which I start virtually everything I write and type, formal and informal. I borrowed it from my cousin Nkrumah Kenyatta Byrd at Indefinite Designs, the clothing company I worked for in the late 1990s and early 2000s. My version grew out of his itGODfaot, but the meaning is the same. The i, which stands for the individual, is lower-cased, representing humility toward our Creators. The rest is self-explanatory: thanking the Creators for everything on this earthly plain that we are fortunate to have the blessing to participate with and in—the good, the bad, and the ugly. This tagline keeps me grounded and humble in all my communications, notes, emails, and doodles as well as centers the Creators in almost every thought I have. It has become rote in my writing, but not in my thoughts.

Thank you to my Creators and my Ancestors, for what you invested in me, in knowledge, in information, in genetic material, and just in existing. My parents, Mr. Michael and Ann Carter Sr., have been my biggest supporters since conception and I'm grateful. My father kept my hands in the soil, and my butt at Carter Farms, more often than I liked as a young person. My mother instilled in me a love for learning. I don't remember my first library visit, but I remember going to the library often, and it being a place I could spend hours exploring, learning, laughing, and keeping myself entertained. The library was the only place that you were guaranteed to leave with something. I couldn't always get the toy I wanted, or the cereal I thought I wanted, but I always left with the books I wanted, and it was always books, never a singular book. Thank you for giving me the gift of inquisitiveness. Thank you to Mrs. Louise "Nana" French—she was my unrelated grandmother—who poured love into me that I felt through her cooking and her words. She would fry a whole chicken just for me, because I loved her cooking, and her positive words and energy has stayed with me to this day. She is the first woman I ever loved who wasn't my mother. My grandmother Mrs. Lucille Carter, whose quiet strength and presence keeps me grounded, and I do miss her sweet potato, white potato, and apple pies, which, once I got old enough, she made me my own; I didn't have to share with noooobody. She loved me so much she veganized them for me to ensure I could still taste her love.

Thank you to Erika Howsare for your work and support in getting this book done. An interview on Gran's couch turned into this; I'm sure neither of us saw that coming. It's been greatly appreciated and valued.

Thank you to my cousin Mrs. Ashleigh Corrin Webb, an award-winning illustrator whom I felt ancestrally compelled to include in this writing. The chapter opening illustrations are the gifts of her mind and pencil. Gran, Aunt Marjorie, Grandma Lucille, and Uncle John are beaming with pride that we were able to collaborate in this way to share an aspect of their story. Thank you, Ashleigh.

Thank you to all the farmers and growers I learned from, I've worked for and with; I'm appreciative of you and your efforts. Farming is the noblest and oldest profession, and it takes a great person to partake in it. Thank you for stepping into your greatness.

Thank you to Lehesmik, Mamaga 4.3, and the residents of M83, you have inspired me in different ways to share my story and get this out.

Thank you to Drs. Ira and Gladys West, I'm forever grateful for your contributions, your intelligence, and your loving relationships with our family.

Thank you to University of Virginia Environmental Thought and Practice department, Dr. Paul Freedman, and Dr. Lisa Shutt for playing your parts to allow me to teach Africulture at the University of Virginia. Thank you to my students, past, present, and future.

My future, my sons, thank you for being you. You have and continue to inspire more than you know. No, I haven't grown a beard, but my love and appreciation for you all has grown immensely while penning this book.

Thank you to those who may be inspired by this book, learn something, grow something, invest in something related to the land, local farmers, Black farmers, Indigenous farmers, Asian farmers, small farmers, beginner farmers, veteran farmers, and specialty crop farmers in general.

Thank you for reading this book.

APPENDIX A

TIMELINE

7,000 years ago: The Kingdom of Kush on the Nile River possibly domesticates cotton

1324: Mansa Musa makes a hajj to Mecca, feeding a caravan of 60,000-plus along the way

1342: Portuguese colonizers arriving in Africa note locals' skill with indigo dye

1455: Pope Nicholas V issues a papal bull permitting the enslavement of Africans

1482: First European outpost is built in West Africa by order of Portugal's King John

1513: Spanish slave ships bring kidnapped Africans to South Carolina

1555: Construction begins on a trading post in Ghana that would become the Cape Coast slave dungeon in the 1600s

1607: Cotton arrives in Virginia with English colonizers (and may have been planted as early as 1556 in the Florida colony)

1613: The Shirley Plantation, the oldest in Virginia and a seat of my family history, is settled

1619: First account of enslaved or indentured Africans arriving on Turtle Island, a.k.a. North America

1640: Mr. John Punch becomes the first Black person to be enslaved in America under criminal law

1600s or 1700s: Some of my ancestors are brought to Virginia and enslaved on plantations

1653: Swedish African Company builds a wooden fort at Cape Coast alongside the trading post, which was first used for timber and gold trading and then for the slave trade of African bodies

1662: The law declares that the child of any enslaved mother in Virginia is also enslaved for life

1685: Rice may have arrived in the South Carolina colony via an errant ship from Madagascar

1700–1720: The rice industry based in South Carolina grows fiftyfold with the help of enslaved labor

1732: Ambrose Madison, grandfather of fourth US President James Madison, dies from being poisoned by three enslaved people: Uncle Pompey, Uncle Turk, and Auntie Dido

1739: The Stono Rebellion occurs in South Carolina as enslaved freedom seekers fight to link with the Seminole in Florida; the following year, the playing of drums is banned in South Carolina

1740: Eliza Lucas Pinckney, an eighteen-year-old plantation owner, establishes the indigo industry with the knowledge and labor of enslaved experts

1745: My ancestors Mr. Angus McIntosh and Mr. Peter McIntosh arrive in Virginia from Scotland

1745: Auntie Eve is accused of poisoning her enslaver; she will be burned to death in 1746

1748: The Virginia Poisoning Act of 1748 is enacted by the Virginia House of Burgesses

1775: At the peak of the South Carolina indigo industry, 1 million pounds are traded to Britain

1793: Eli Whitney invents the cotton gin

1800: On August 24, an enslaved Virginia man gathers a group of freedom fighters to fight for their freedom by capturing Virginia governor James Monroe

1811: On January 8, Mr. Charles Deslondes leads an enslaved uprising in territory that later became part of the state of Louisiana; the largest such uprising of enslaved people for freedom in American history

1820: North Carolina House Representative Lewis Williams sponsors a resolution to create the US House Committee on Agriculture

1822: American Colonization Society (ACS) establishes a settlement for freed Black men and Black women in territory now part of present-day Liberia

1829: David Walker writes an anti-slavery document, *Appeal to the Colored Citizens of the World*, in Boston, which is circulated throughout the United States

1834: Mr. Henry Blair, a free Black farmer in Maryland, invents the mechanical corn planter

1841: Mr. Edmond Albius discovers a method for pollinating vanilla

1850: Cotton is by far America's main export at 2.85 million bales

1857: Roger Taney's opinion in the *Dred Scott v. Stanford* decision denies citizenship to Black Americans

1859: Possible founding of Freetown by my fifth great-grandfather, Mr. Robert "Punch" Ellis

1861: The Corwin Amendment, designed to protect the institution of slavery from interference by federal power, is approved by the House, the Senate, and President James Buchanan (but was subsequently never ratified)

1862: President Lincoln recognizes Liberia, opening the coffee trade

1862: Establishment of the USDA and the land-grant (or "land-grab") universities

1863: The Emancipation Proclamation is issued

1865: General William Tecumseh Sherman issues Special Field Order No. 15 to redistribute land to formerly enslaved families

1865: Ratification of the Thirteenth Amendment, abolishing slavery

1866: The first Civil Rights Act is passed

1875–1955: More than 4,700 people are lynched during Reconstruction and American apartheid

1877: Federal troops are pulled from the South

1879: Mr. Tetteh Quarshie founds the cocoa industry in Ghana

1870s: Negative stereotypes about watermelon begin to proliferate

1881: Dr. Booker T. Washington becomes the first principal of Tuskegee Institute

1890: Establishment of agricultural schools specifically for Black students ("HBCUs")

1896: Dr. George Washington Carver joins the faculty at Tuskegee University

1900: Virginia has more Black farmers than any state in America, at over 25,000

1910: Black people own 925,000 farms, land equivalent to the area of South Carolina

1910: My great-great-grandparents Mr. Jefferson Davis Shirley and Mrs. Catherine Walker Shirley purchase 150 acres in Orange County, VA, for $722.05

1914: Smith-Lever Act creates a nationalized extension service that will mostly benefit white farmers

1915: My great-grandfather Mr. John Lewis Carter purchases 25 acres on what is now Shirley Road

1916: Federal Farm Loan Act establishes funding source that will mostly benefit white farmers

1920: Black farmers own over 15 million acres of farmland

1925: Virginia continues to lead the country in number of Black farmers, now with 50,000

1931: A lien is put in place on the Carter Farms property

1931: Agricultural Adjustment Act enacted

1935: National Labor Relations Act of 1935 protects unions but excludes farm workers, as do the 1935 Social Security Act and the 1938 Fair Labor Standards Act

1937: Bankhead-Jones Farm Tenant Act of 1937 offers support to tenant farmers but mostly excludes Black farmers

1945: My great-grandmother moves her family to Carters Lane after paying off the land

1968: My uncle Alphonso dies in a tractor accident

1972: HBCUs finally receive federal funding for extension services

1980s: Farm Bills cut food stamp allotments, affecting urban Black communities

1985: "We Are the World" perpetuates a perception of Africa as monolithic and desperate

1980s and 1990s: African consciousness hits a peak in hip-hop culture

1994: The USDA finds that minorities have been discriminated against in crop payments, disaster payments, and loans

1997: The *Pigford v. Glickman* class-action suit addresses racial discrimination by the USDA

1999: *Pigford v. Glickman* is settled by Judge Paul L. Friedman, who orders the USDA to make payments of more than $2 billion

2010: A second round of settlements, called Pigford II and amounting to $1.2 billion, is negotiated

2012: My family and I arrive in Ghana

2014: Congress publicly acknowledges discriminatory funding for extension services

2017: After five years in Ghana, I return to Virginia and officially found Carter Farms

2017: Virginia is home to only about 1,300 Black farmers

2017–2022: Virginia loses another 20 percent of our Black farmers (and 10 percent of our farmers overall)

2020: Africulture becomes an official nonprofit

2020: Mr. George Floyd is murdered

2021: I am invited to help University of Virginia dining increase engagement with socially disadvantaged and local farmers

2021: Equity Commission is established at the USDA along with a $4 billion fund for Black farmers through the American Rescue Plan

2022: America finally gets a federal anti-lynching law

2025: The forty-seventh president of the United States issues an executive order to terminate funding for equity-related programs in the USDA and other agencies

2025: The USDA states it will end preferences based on race and sex in response to the *Strickland vs. USDA* court ruling

APPENDIX B

NUTRIENT LEVELS IN CROPS AT CARTER FARMS

In 2023, I hired American Climate Partners to take leaf tissue from various plants grown on our farm for macronutrient analysis at Waypoint Analytical Laboratories. Leaf tissue analysis shows exactly what nutrients the plant

Table C.1. Nutrient Levels in Crops at Carter Farms

	Amaranth	Jewels of Opar	Nigerian spinach
Nitrogen (3.00–3.99)*	3.15	3.36	3.16
Sulfur (0.20–0.59)*	0.32	0.36	0.29
Phosphorus (0.20–0.29)*	0.26	0.16	0.21
Potassium (2.00–3.99)*	4.63	7.29	4.19
Magnesium (0.25–0.49)*	1.07	2.58	1.32
Calcium (0.80–1.59)*	2.64	2.07	2.50
Boron (20–50)**	31	17	25
Manganese (40–100)**	1196	1115	1341
Zinc (25–50)**	116	95	140
Iron (40–251)**	106	114	91
Copper (5–10)**	14	10	19

* normal range of values (% dry weight)

** normal range of values (parts per million)

has extracted from the soil and spread throughout their stems and leaves. I've found it a more effective approach to understanding plant nutrition than just a soil test, which analyzes what nutrients are in the soil.

Sweet potato	Taro leaf/ Elephant ear	Collards	Kale	Lettuce
3.59	3.40	3.28	4.13	3.59
0.39	0.28	0.78	0.59	0.35
0.26	0.32	0.21	0.19	0.33
3.89	3.96	4.04	5.10	6.43
0.82	0.60	0.58	0.67	0.50
1.50	1.80	2.19	2.64	1.16
35	23	25	16	23
406	1285	108	395	290
27	60	78	99	93
112	93	115	191	161
16	13	7	7	15

APPENDIX C

THE CARTER FAMILY TREE

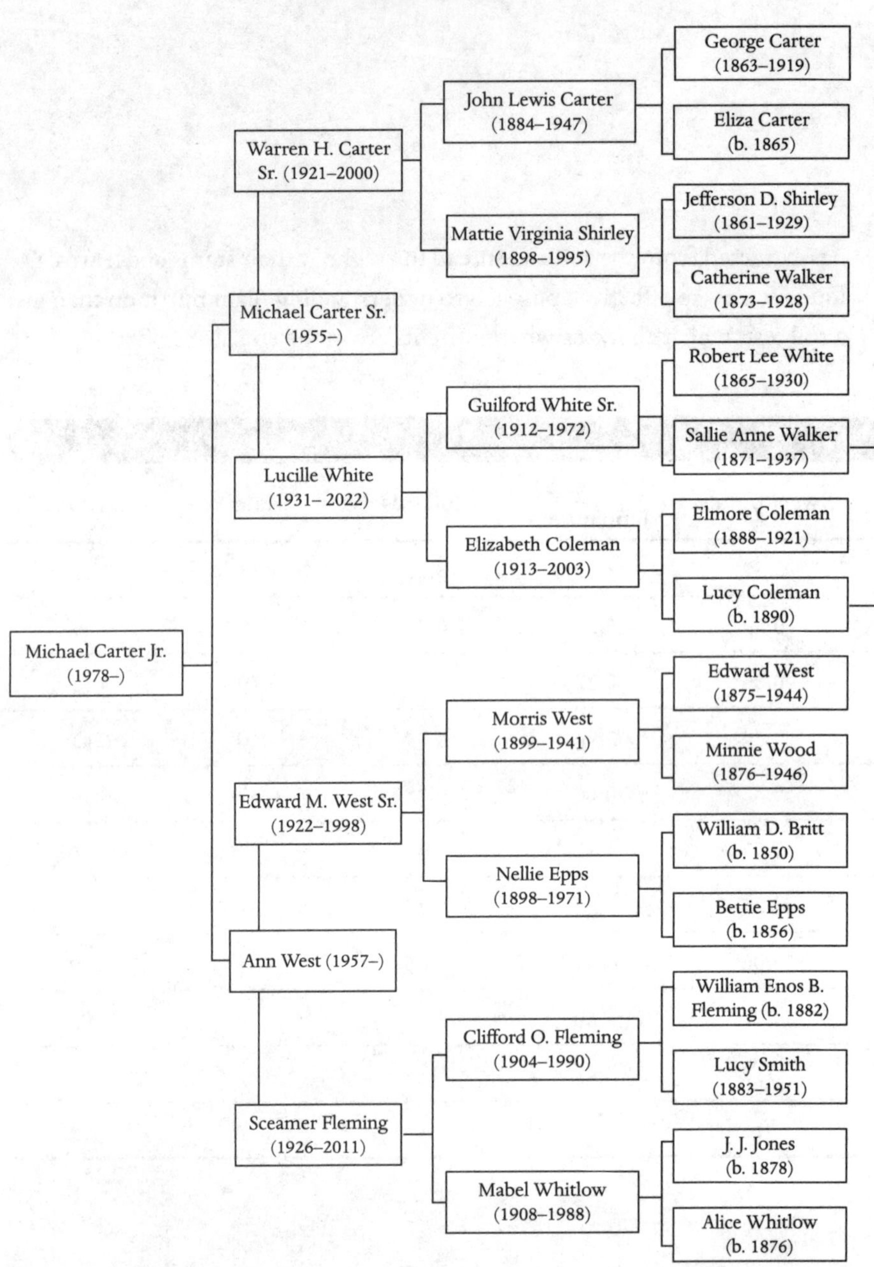

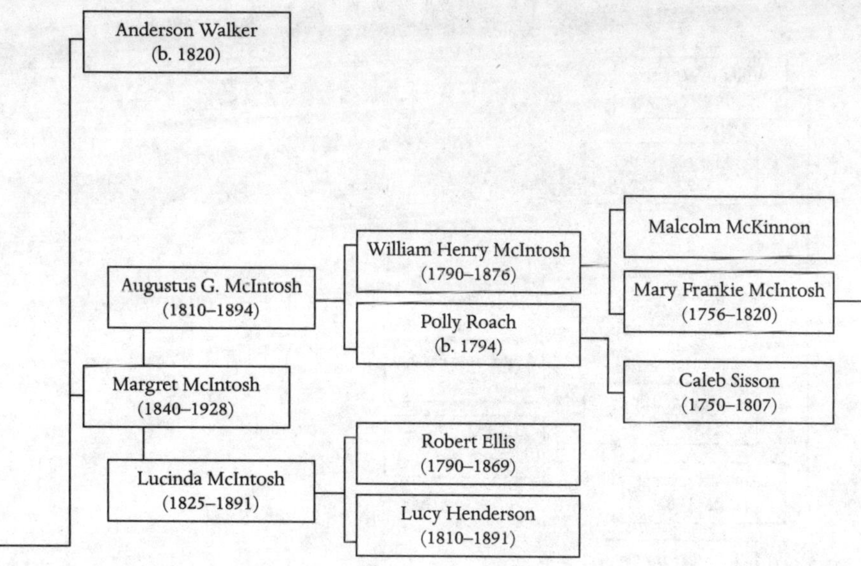
Anderson Walker
(b. 1820)
Augustus G. McIntosh
(1810–1894)
Margret McIntosh
(1840–1928)
Lucinda McIntosh
(1825–1891)
William Henry McIntosh
(1790–1876)
Polly Roach
(b. 1794)
Robert Ellis
(1790–1869)
Lucy Henderson
(1810–1891)
Malcolm McKinnon
Mary Frankie McIntosh
(1756–1820)
Caleb Sisson
(1750–1807)

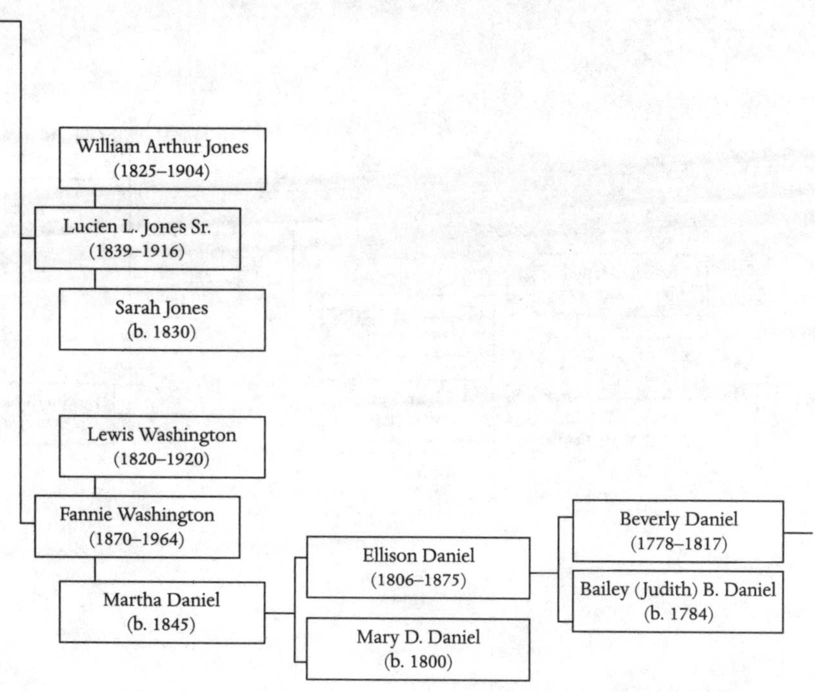
William Arthur Jones
(1825–1904)
Lucien L. Jones Sr.
(1839–1916)
Sarah Jones
(b. 1830)
Lewis Washington
(1820–1920)
Fannie Washington
(1870–1964)
Martha Daniel
(b. 1845)
Ellison Daniel
(1806–1875)
Mary D. Daniel
(b. 1800)
Beverly Daniel
(1778–1817)
Bailey (Judith) B. Daniel
(b. 1784)

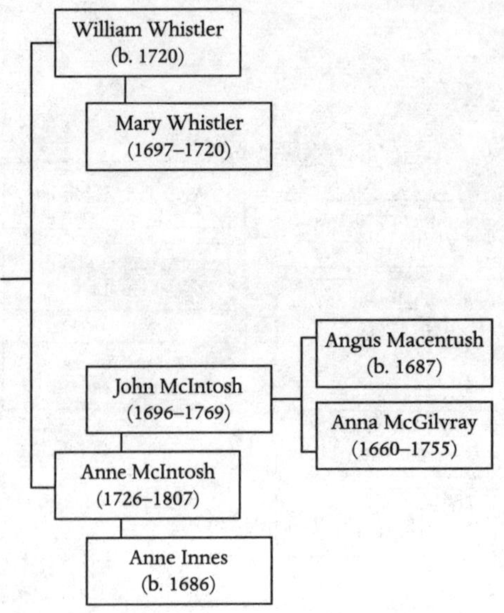
William Whistler
(b. 1720)
Mary Whistler
(1697–1720)
John McIntosh
(1696–1769)
Angus Macentush
(b. 1687)
Anna McGilvray
(1660–1755)
Anne McIntosh
(1726–1807)
Anne Innes
(b. 1686)

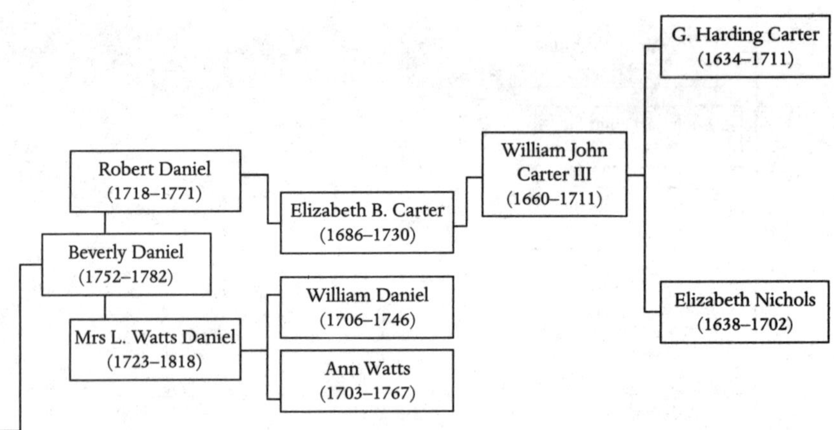
G. Harding Carter
(1634–1711)
William John
Carter III
(1660–1711)
Robert Daniel
(1718–1771)
Elizabeth B. Carter
(1686–1730)
Beverly Daniel
(1752–1782)
William Daniel
(1706–1746)
Elizabeth Nichols
(1638–1702)
Mrs L. Watts Daniel
(1723–1818)
Ann Watts
(1703–1767)

NOTES

Chapter 1. Seeds

1. Howard Zinn, *A People's History of the United States: 1492–Present* (HarperCollins, 2003), 24.

Chapter 2. Soil

1. "George Washington Carver: A National Agricultural Library Digital Exhibit," National Agricultural Library, accessed August 22, 2025, https://www.nal.usda.gov/exhibits/ipd/carver/exhibits/show/exhibits/carver-exhibits.
2. "USDA Soil Taxonomy," Wikipedia, last modified June 28, 2025, 00:36, https://en.wikipedia.org/w/index.php?title=USDA_soil_taxonomy&oldid=1297710761.
3. "Sankofa Meaning: What Is the Ghanian Concept About?," MasterClass, October 20, 2022, https://www.masterclass.com/articles/sankofa-meaning-explained.
4. Bruce J. Reynolds, *Black Farmers in America, 1865–2000: The Pursuit of Independent Farming and the Role of Cooperatives*, RBS Research Report no. 194 (US Department of Agriculture Rural Business Cooperative Service, 2002), https://www.rd.usda.gov/files/RR194.pdf.
5. Denise A. Smith, "Nourishing the Nation While Starving: The Underfunding of Black Land-Grant Colleges and Universities," The Century Foundation, July 24, 2023, accessed September 25, 2025, https://tcf.org/content/report/nourishing-the-nation-while-starving-the-underfunding-of-black-land-grant-colleges-and-universities.
6. *Hearing to Review the Smith-Lever Act on Its 100th Anniversary: Hearing Before the Subcommittee on Horticulture, Research, Biotechnology, and Foreign Agriculture of the Committee on Agriculture*, US Congress, House of Representatives, 113th Cong., 2nd session, March 4, 2014 (testimony of Dr. L. Washington Lyons), https://www.govinfo.gov/content/pkg/CHRG-113hhrg87047/html/CHRG-113hhrg87047.htm.
7. Budget Bill - HB 1800 (Introduced), Virginia House of Delegates, 2021 Session, https://budget.lis.virginia.gov/item/2021/1/HB1800/Introduced/1/241; Budget Bill - HB 30 (Introduced), Virginia House of Delegates, 2024 Session, https://budget.lis.virginia.gov/item/2024/1/HB30/Introduced/1/221; Budget Bill - HB 29 (Chapter 1072), Virginia House of Delegates, 2000 Session, https://budget.lis.virginia.gov/item/2000/1/HB29/Chapter/1/256.

8. Servicemen's Readjustment Act of 1944, US Congress, House, Public Law 346, 78th Cong., 2nd sess., June 22, 1944, https://vetsedsuccess.org/wp-content/uploads/2018/09/jun-22-1944-gi-bill.pdf.
9. Catharine Paddock, "Soil Bacteria Work in Similar Way to Antidepressants," Medical News Today, April 2, 2007, https://www.medicalnewstoday.com/articles/66840; Balachundhar Subramaniam, "The Surprising Ways Soil Can Aid Our Mental Health," Psychology Today, August 29, 2024, https://www.psychologytoday.com/us/blog/a-subtle-impact/202408/what-does-soil-have-to-do-with-mental-health.
10. Lisa Marshall, "Why Dirt May Be Nature's Original Stress-Buster," *CU Boulder Today*, May 9, 2019, https://www.colorado.edu/today/2019/05/09/natures-original-stress-buster.

Chapter 3. Roots

1. Chris Duvall, *The African Roots of Marijuana* (Duke University Press, 2019), 24.
2. Orange County, Virginia, Deed Book, Circuit Court Clerk's Record Room, 367–68.
3. William Waller Hening, ed., *The Statutes at Large: Being a Collection of All the Laws of Virginia, from the First Session of the Legislature, in the Year 1619: Published Pursuant to an Act of the General Assembly of Virginia, Passed on the Fifth Day of February One Thousand Eight Hundred and Eight*, Vol. II (Printed by and for Samuel Pleasants, junior, printer to the commonwealth, 1809), 170.
4. Jack Frazer, "The (Almost) Lahore Railroad Line," *Orange County Historical Society Record* 50, no. 2 (Fall 2019), 1–3, https://www.orangecovahist.org/_files/ugd/33a0c8_778a363354b44f149fc583d212da5054.pdf.
5. Bill Ruehlmann, "The Queen of Regional Cuisine," *The Virginian-Pilot* (Hampton Roads, VA), May 26, 1996, https://scholar.lib.vt.edu/VA-news/VA-Pilot/issues/1996/vp960526/05230056.htm.
6. Dred Scott v. John F. A. Sandford, 60 U.S. 393 (1857), National Archives, https://www.archives.gov/milestone-documents/dred-scott-v-sandford.
7. "Selected Farm Characteristics by Race: 2022 and 2017," 2022 Census of Agriculture, US Department of Agriculture, Vol 1 Geographic Area Series, Part 51, table 61, https://www.nass.usda.gov/Publications/AgCensus/2022/Full_Report/Volume_1,_Chapter_1_US/st99_1_061_061.pdf.

Chapter 4. Compost

1. Ebonie Alexander and Parker Agelasto, "Ensuring Rightful Property Ownership Through the Uniform Partition of Heirs Property Act," Virginia Conservation Network, accessed September 27, 2025, https://www.vcnva.org/wp-content/uploads/2019/08/HEIRS-PROPERTY.pdf.
2. June Purcell Guild, *Black Laws of Virginia* (Whittet & Shepperson, 1936; Heritage Books, 2011), 21, 37. Citation refers to the Heritage Books edition.

3. Carl Linnaeus, *Linnaeus's Philosophia Botanica* (1751), 171, quoted in William Whewell, *History of the Inductive Sciences: From the Earliest to the Present Times*, vol. 3 (D. Appleton & Co, 1870), 355–56.
4. Isabelle Charmantier, "Linnaeus and Race," The Linnean Society, September 3, 2020, https://www.linnean.org/learning/who-was-linnaeus/linnaeus-and-race.
5. "Cheese Caves and Food Surpluses: Why the U.S. Government Currently Stores 1.4 Billion lbs of Cheese," The Farmlink Project, August 19, 2021, https://www.farmlinkproject.org/stories-and-features/cheese-caves-and-food-surpluses-why-the-u-s-government-currently-stores-1-4-billion-lbs-of-cheese.

Chapter 5. First Leaves

1. Michael W. Twitty, "Grain of Truth," *Edge Magazine*, accessed September 29, 2025, https://edgemagonline.com/article/grain-of-truth.
2. Jude C. Aguwa, "Culture Summary: Mende," in *eHRAF World Cultures* (Human Relations Area Files, 2010), accessed September 29, 2025, https://ehrafworldcultures.yale.edu/cultures/fc07/summary.
3. Steven Linscombe, "The History of U.S. Rice Production - Part 1," LSU Agricultural Center, December 8, 2006, https://www.lsuagcenter.com/portals/our_offices/research_stations/rice/features/publications/the-history-of-us-rice-production--part-1.
4. "Rice Production in the United States," Wikipedia, last modified September 7, 2025, 16:05, https://en.wikipedia.org/wiki/Rice_production_in_the_United_States.
5. Judith Carney, "Rice Cultivation in the History of Slavery," in *Oxford Research Encyclopedia of African History* (Oxford University Press, June 2020), https://doi.org/10.1093/acrefore/9780190277734.013.713.
6. "U.S. Rice Market Size & Outlook, 2021–2028," Grand View Horizon Databooks, accessed September 29, 2025, https://www.grandviewresearch.com/horizon/outlook/rice-market/united-states.
7. "History of Cotton," Wikipedia, last modified September 25, 2025, 15:19, https://en.wikipedia.org/wiki/History_of_cotton.
8. "Cotton," Wikipedia, last modified September 8, 2025, 21:39, https://en.wikipedia.org/wiki/Cotton.
9. "The Story of Cotton," National Cotton Council of America, accessed September 29, 2025, https://www.cotton.org/pubs/cottoncounts/story.
10. "Henry Blair (inventor)," Wikipedia, last modified September 25, 2025, 22:49, https://en.wikipedia.org/wiki/Henry_Blair_(inventor).
11. Howard Zinn, *A People's History of the United States: 1492–Present* (HarperCollins, 2003), 25.
12. J. David Hacker, "From '20. and Odd' to 10 Million: The Growth of the Slave Population in the United States," *Slavery & Abolition* 41, no. 4 (2020): 840–55, https://doi.org/10.1080/0144039x.2020.1755502.

13. "Harrison Ruffin Tyler, Preserver of Virginia History and Grandson of 10th US President, Dies at 96," *Associated Press*, May 29, 2025, https://apnews.com/article/president-john-tyler-grandson-death-virginia-dc21fc685f61e3fd2f0db2ad110a2f7a; Sarah Kuta, "Last Surviving Grandson of President John Tyler, Who Took Office in 1841, Dies at 96," *Smithsonian Magazine*, May 30, 2025, https://www.smithsonianmag.com/smart-news/last-surviving-grandson-of-president-john-tyler-who-took-office-in-1841-dies-at-96-180986724.
14. "Each Person In America Owns an Average of 7 Pairs of Jeans," *South Florida Reporter*, December 5, 2024, https://southfloridareporter.com/each-person-in-america-owns-an-average-of-7-pairs-of-jeans.
15. Latria Graham, "The Blue That Enchanted the World," *Smithsonian Magazine*, November/December 2022, https://www.smithsonianmag.com/arts-culture/indigo-making-comeback-south-carolina-180980987.
16. "Indigo Dye," Wikipedia, last modified September 1, 2025, 23:07, https://en.wikipedia.org/wiki/Indigo_dye; Catherine E. McKinley, "Indigo: The Indelible Color That Ruled The World," interview by Michel Martin, *Tell Me More*, NPR, November 7, 2011; Catherine E. McKinley, "Prologue," from *Indigo: In Search of the Color That Seduced the World* (Bloomsbury, 2011), excerpted on *NPR*, November 7, 2011, https://www.npr.org/2011/11/07/142094495/excerpt-indigo-in-search-of-the-color-that-seduced-the-world.
17. Imani Perry, *Black in Blues: How a Color Tells the Story of My People* (Ecco, 2025), 34.
18. Nic Butler, "Indigo in the Fabric of Early South Carolina," *Charleston Time Machine*, Charleston County Public Library, August 16, 2019, https://www.ccpl.org/charleston-time-machine/indigo-fabric-early-south-carolina.
19. Jesslyn Shields, "The Dark History of Indigo, Slavery's Other Cash Crop," HowStuffWorks, February 7, 2020, https://history.howstuffworks.com/world-history/indigo.htm.
20. Lilith Dorsey, "Hoodoo How We Do: Indigo," *Voodoo Universe*, Patheos, September 19, 2024, https://www.patheos.com/blogs/voodoouniverse/2024/09/hoodoo-how-we-do-indigo; Anitra Simone, "Elukami Indigo Fingers," Urban Dictionary, August 5, 2025, https://www.urbandictionary.com/define.php?term=Elukami+Indigo+Fingers.
21. "George Washington Carver," Britannica, last modified September 10, 2025, https://www.britannica.com/biography/George-Washington-Carver.
22. Camila Domonoske, "A Legume with Many Names: The Story of 'Goober,'" *Code Switch*, NPR, April 20, 2014, https://www.npr.org/sections/codeswitch/2014/04/20/304585019/a-legume-with-many-names-the-story-of-goober; Jori Lewis, *Slaves for Peanuts: A Story of Conquest, Liberation, and a Crop That Changed History* (The New Press, 2022), 23–28.
23. "The Legacy of Dr. George Washington Carver," Tuskegee University, accessed August 22, 2025, https://www.tuskegee.edu/legacy/george-washington-carver.html.

24. "Arts and Crafts," George Washington Carver Museum Management Program, National Park Service, accessed October 27, 2025, https://www.nps.gov/museum/exhibits/tuskegee/gwcarver/arts_crafts.html, page has been removed from the National Park Service website as of November 12, 2025.
25. Carver to Hubert W. Pelt, February 24, 1930, in *George Washington Carver: In His Own Words*, ed. G. R. Kremer (University of Missouri Press, 1987), 142-3, https://www.vaticanobservatory.org/wp-content/uploads/2018/01/How-to-Search-for-the-Truth.pdf.
26. Tim Harford, "How Rudolf Diesel's Engine Changed the World," *50 Things That Made the Modern Economy*, BBC News, December 19, 2016, https://www.bbc.com/news/business-38302874.
27. "Carver, Ford Took Early Lead in Biofuel Development," Reliable Plant, accessed September 30, 2025, https://www.reliableplant.com/Read/5513/carver-ford-biofuel.
28. "Booker T. Washington," History, last modified May 28, 2025, https://www.history.com/articles/booker-t-washington.
29. Booker T. Washington, *Up From Slavery: An Autobiography* (A. L. Burt, 1901), 4.
30. Mary Kate Eckles, "W. E. B. Du Bois, the Bureau of Labor Statistics, and the Study of Black Life," Rediscovering Black History, National Archives, June 30, 2015, https://rediscovering-black-history.blogs.archives.gov/2015/06/30/w-e-b-du-bois-the-bureau-of-labor-statistics-and-the-study-of-black-life.
31. "Tetteh Quarshie," Wikipedia, last modified June 15, 2025, 15:40, https://en.wikipedia.org/wiki/Tetteh_Quarshie.
32. "Cocoa Producing Countries 2025," World Population Review, accessed September 30, 2025, https://worldpopulationreview.com/country-rankings/cocoa-producing-countries.
33. "Vanilla," Wikipedia, last modified September 20, 2025, 13:48, https://en.wikipedia.org/wiki/Vanilla.
34. "Edmond Albius," Wikipedia, last modified September 17, 2025, 07_05, https://en.wikipedia.org/wiki/Edmond_Albius.
35. Simran Sethi, "The Bittersweet Story of Vanilla," *Smithsonian Magazine*, April 3, 2017, https://www.smithsonianmag.com/science-nature/bittersweet-story-vanilla-180962757.
36. Leanne Melbourne, "Edmond Albius: The Boy Who Revolutionised the Vanilla Industry," The Linnean Society, October 16, 2019, https://www.linnean.org/news/2019/10/16/edmond-albius.
37. "History of Coffee," Wikipedia, last modified September 20, 2025, 21:47, https://en.wikipedia.org/wiki/History_of_coffee.
38. "Coffee - United States," Statista, accessed September 30, 2025, https://www.statista.com/outlook/cmo/hot-drinks/coffee/united-states.
39. Bronwen Everill, "How Coffee Helped the Union Caffeinate Their Way to Victory in the Civil War," *Smithsonian Magazine*, July/August 2024, https://

www.smithsonianmag.com/history/how-coffee-helped-the-union-caffeinate-their-way-victory-civil-war-180984502.

40. Everill, "How Coffee Helped"; Ashley Webb, "Coffee and the Civil War Soldier," American Battlefield Trust, last modified November 19, 2024, https://www.battlefields.org/learn/articles/coffee-and-civil-war-soldier.
41. Judith A. Carney and Richard Nicholas Rosomoff, *In the Shadow of Slavery: Africa's Botanical Legacy in the Atlantic World* (University of California Press, 2009), 70.

Chapter 6. True Leaves

1. Charles Walter, *Minerals for the Genetic Code: An Exposition & Analysis of the Dr. Olree Standard Genetic Periodic Chart & the Physical, Chemical & Biological Connection* (Acres USA, 2024), 117.
2. A. G. Grinnan, "The Burning of Eve in Virginia," in "Historical Notes and Queries," *The Virginia Magazine of History and Biography* 3 (1896): 308–10, https://www.jstor.org/stable/4241905.
3. Grinnan, "The Burning of Eve in Virginia."
4. Ann L. Miller, *The Short Life and Strange Death of Ambrose Madison* (Orange County Historical Society, 2001), 26–27.
5. Stefano Mancuso, *The Revolutionary Genius of Plants: A New Understanding of Plant Intelligence and Behavior* (Atria, 2018), 6–11.
6. National Research Council, *Lost Crops of Africa: Volume II: Vegetables* (The National Academies Press, 1996), https://doi.org/10.17226/11763, 35, 37–38.
7. National Research Council, *Lost Crops of Africa: Volume II*, 35.
8. Anu Rastogi and Sudhir Shukla, "Amaranth: A New Millennium Crop of Nutraceutical Values," *Critical Reviews in Food Science and Nutrition* 53, no. 2 (2013): 109–25.
9. Deirdre Campbell et al., "Hibiscus Tea, Hormone Balance, and Thrombosis: A Case Report," *Integrative Medicine: A Clinician's Journal* 22, no. 2 (2023): 40–42.

Chapter 7. Pestilence and Resistance

1. James H. Sweet, "The Iberian Roots of American Racist Thought," *The William and Mary Quarterly* 54, no. 1 (January 1997): 143–66.
2. Claud Anderson, *PowerNomics: The National Plan to Empower Black America* (PowerNomics Corp. of America, 2000), 5.
3. Leo Wiener, *Africa and the Discovery of America*, Volume 1 (Innes & Sons, 1920), 34.
4. Howard Zinn, *A People's History of the United States: 1492–Present* (HarperCollins, 2003), 7.
5. James H. Sweet, "Spanish and Portuguese Influences on Racial Slavery in British North America, 1492–1619" (paper presented at the Fifth Annual Gilder Lehrman Center International Conference, "Collective Degradation: Slavery and the

Construction of Race," Yale University, New Haven, CT, November 7–8, 2003), 7, https://macmillan.yale.edu/sites/default/files/files/events/race/Sweet.pdf.

6. William Loren Katz, "The Forgotten First Attempt to Plant a Colony on US Soil," History News Network, January 29, 2017, adapted from *Black Indians: A Hidden Heritage* (Atheneum, 2014 revised edition), https://www.historynewsnetwork.org/article/the-forgotten-first-attempt-to-plant-a-colony-on-u.
7. Katz, "The Forgotten First Attempt to Plant a Colony on US Soil."
8. H. R. McIlwane, ed., *Minutes of the Council and General Court of Colonial Virginia 1622–1632, 1670–1676* (Library of Virginia, 1924), 466–67.
9. W. E. B. Du Bois, *Black Reconstruction in America (The Oxford W. E. B. Du Bois): An Essay Toward a History of the Part Which Black Folk Played in the Attempt to Reconstruct Democracy in America, 1860–1880* (Oxford University Press, 2007), 700.
10. Thomas Jefferson, Notes on the State of Virginia (J. W. Randolph, 1853), 155.
11. Dred Scott v. John F. A. Sandford, 60 U.S. 393 (1857), National Archives, https://www.archives.gov/milestone-documents/dred-scott-v-sandford.
12. "Homesteading by the Numbers," Homestead National Historical Park, Nebraska, National Park Service, last modified March 17, 2025, https://www.nps.gov/home/learn/historyculture/bynumbers.htm.
13. Andrew Muhammad, Christopher Sichko, and Tore C. Olsson, "African Americans and Federal Land Policy: Exploring the Homestead Acts of 1862 and 1866," *Applied Economic Perspectives and Policy* 46, no. 1 (March 2024): 95–110, https://doi.org/10.1002/aepp.13401.
14. "African American Homesteaders in the Great Plains," National Park Service, last modified March 7, 2023, https://www.nps.gov/articles/african-american-homesteaders-in-the-great-plains.htm.
15. Precious Tshabalala, "A Brief History of Discrimination Against Black Farmers—Including by the USDA." *The Equation* (blog), Union of Concerned Scientists, August 27, 2024. Accessed August 22, 2025, https://blog.ucs.org/precious-tshabalala/a-brief-history-of-discrimination-against-black-farmers-including-by-the-usda, post has been removed from *The Equation* as of October 8, 2025.
16. "United States Colored Troops History," African American Civil War Memorial Museum, accessed October 31, 2025, https://afroamcivilwar.org/united-states-colored-troops-history.
17. Nick Overby, "Supplying Hell: The Campaign for Atlanta," from the *Quartermaster Professional Bulletin* (Winter 1992), Army Quartermaster Foundation, accessed October 31, 2025, https://quartermasterfoundation.org/supplying-hell-the-campaign-for-atlanta; "Atlanta Campaign," Wikipedia, last modified August 19, 2025, 05:16, https://en.wikipedia.org/wiki/Atlanta_campaign; "Marker Monday: African-American Soldiers in Combat," Georgia Historical Society, accessed October 31, 2025, https://www.georgiahistory.com/marker-monday-african-american-soldiers-in-combat.

18. "Racial Discrimination in Union Army Pensions Detailed by New Study," Brigham Young University Communications, News, February 10, 2010, https://news.byu.edu/news/racial-discrimination-union-army-pensions-detailed-new-study.
19. Tom Shoop, "Harriet Tubman's Three-Decade Battle for a Federal Pension," *Government Executive*, June 17, 2021, https://www.govexec.com/pay-benefits/2021/06/harriet-tubmans-three-decade-battle-federal-pension/174812.
20. Tunde Adeleke, *In the Service of God and Humanity: Conscience, Reason, and the Mind of Martin R. Delany* (University of South Carolina Press, 2021), 60, https://doi.org/10.48172/9781643361857.
21. "Civil Rights Act of 1866, 'An Act to Protect All Persons in the United States in Their Civil Rights, and Furnish the Means of Their Vindication,'" National Constitution Center, accessed September 25, 2025, https://constitutioncenter.org/the-constitution/historic-document-library/detail/civil-rights-act-of-1866-april-9-1866-an-act-to-protect-all-persons-in-the-united-states-in-their-civil-rights-and-furnish-the-means-of-their-vindication.
22. "Freedman's Bank Demise," United States Department of the Treasury, accessed August 22, 2025, https://home.treasury.gov/about/history/freedmans-bank-building/freedmans-bank-demise.
23. W. E. B. Du Bois, *The Souls of Black Folk* (A. C. McClurg & Co., 1904), 30.
24. *Lynching in America: Confronting the Legacy of Racial Terror*, third ed., Equal Justice Initiative, 2017, accessed September 30, 2025, https://lynchinginamerica.eji.org/report.
25. Angel Harris, "Replacing the Noose With a Needle: The Legacy of Lynching in the United States," *News & Commentary* (blog), American Civil Liberties Union, February 18, 2015, https://www.aclu.org/news/capital-punishment/replacing-noose-needle-legacy-lynching-united-states.
26. David Love, "From 15 Million Acres to 1 Million: How Black People Lost Their Land," America's Black Holocaust Museum, July 2, 2017, https://www.abhmuseum.org/from-15-million-acres-to-1-million-how-black-people-lost-their-land.
27. "Thirty Years of Farmworker Struggle," in "The Road to Sacramento: Marching for Justice in the Fields," article series, National Park Service, accessed October 1, 2025, https://www.nps.gov/articles/000/a-new-era-of-farm-worker-organizing.htm.
28. Tshabalala, "A Brief History of Discrimination Against Black Farmers."
29. Dania V. Francis et al, "Black Land Loss: 1920–1997." *American Economic Association Papers and Proceedings* 112 (May 2022): 38–42, https://doi.org/10.1257/pandp.20221015.
30. Talia Ogliore, "A Seedy Slice of History: Watermelons Actually Came from Northeast Africa," *The Source* (blog), Washington University, May 24, 2021, https://source.washu.edu/2021/05/a-seedy-slice-of-history-watermelons-actually-came-from-northeast-africa.

31. "National Security Study Memorandum 200," Wikipedia, last modified August 11, 2025, 04:12, https://en.wikipedia.org/wiki/National_Security_Study_Memorandum_200.
32. Food Security Act of 1985, H.R. 2100, 99th Cong. (1985), https://www.congress.gov/bill/99th-congress/house-bill/2100.
33. Spencer Rich, "Food Stamp, Child Nutrition Programs Face $2.8 Billion in New Budget Cuts," *The Washington Post*, December 25, 1981, https://www.washingtonpost.com/archive/politics/1981/12/25/food-stamp-child-nutrition-programs-face-28-billion-in-new-budget-cuts/36bda03b-b128-49fa-b601-1d0d0f85090b; Josh Levin, "The True Story behind the 'Welfare Queen' Stereotype," interview by Hari Sreenivasan, *PBS NewsHour*, PBS, June 1, 2019, https://www.pbs.org/newshour/show/the-true-story-behind-the-welfare-queen-stereotype; Martin Gilens, "How the Poor Became Black: The Racialization of American Poverty in the Mass Media," chap. 4 in *Race and the Politics of Welfare Reform*, ed. Sanford F. Schram, Joe Brian Soss, and Richard Carl Fording (University of Michigan Press, 2003), https://press.umich.edu/pdf/9780472068319-ch4.pdf.
34. Sara Breselor, "The Pigford Case: Justice for Black Farmers on Hold," Salon, April 8, 2010, https://www.salon.com/2010/04/08/john_boyd_pigford_glickman_settlement.
35. Tadlock Cowan and Jody Feder, "The *Pigford* Cases: USDA Settlement of Discrimination Suits by Black Farmers," CRS Report for Congress, Congressional Research Service, May 29, 2013, https://nationalaglawcenter.org/wp-content/uploads/assets/crs/RS20430.pdf.
36. "*Pigford v. Glickman*," Wikipedia, last modified December 23, 2024, 13:21, https://en.wikipedia.org/wiki/Pigford_v._Glickman.
37. Cowan and Feder, "The *Pigford* Cases."
38. Wesley Brown, "How the Long Shadow of Racism at USDA Impacts Black Farmers in Arkansas—and Beyond," *Civil Eats*, March 8, 2023, https://civileats.com/2023/03/08/how-the-long-shadow-of-racism-at-usda-impacts-black-farmers-in-arkansas-and-beyond.
39. "Removal of Unconstitutional Preferences Based on Race and Sex in Response to Court Ruling," *Federal Register* 90, no. 130 (July 10, 2025), https://www.federalregister.gov/documents/2025/07/10/2025-12877/removal-of-unconstitutional-preferences-based-on-race-and-sex-in-response-to-court-ruling.
40. "Black Farmers Take Mule, Tractors to D.C.," *ABC News*, August 22, 2002, https://abcnews.go.com/US/story?id=91349.
41. "Removal of Unconstitutional Preferences," *Federal Register*.

Chapter 8. Flowering

1. "FBI COINTELPRO 1967 'Messiah' Memo on 'Black Nationalist Hate Groups,'" *Nation of Islam Research Group*, December 14, 2024, https://noirg.org/document/fbi-cointelpro-1967-messiah-memo-on-black-nationalist-hate-groups.

2. National Agricultural Statistics Service, "Family-Owned Farms Account for 96% of U.S. Farms, According to the Census of Agriculture Typology Report: Small Family Farms Make Up 88% of All U.S. Farms," News Release, United States Department of Agriculture, January 22, 2021, https://www.nass.usda.gov/Newsroom/archive/2021/01-22-2021.php.
3. "Marcus Garvey: 'Look for Me in the Whirlwind', Freedom Speech – (circa) 1924," *The Universal Negro Improvement Association*, June 25, 2020, https://unia-aclgovernment.com/marcus-garvey-look-for-me-in-the-whirlwind-freedom-speech-circa-1924.
4. Jane Dailey, "The Confederate General Who Was Erased," *HuffPost*, August 21, 2017, https://www.huffpost.com/entry/the-confederate-general-who-was-erased-from-history_b_599b3747e4b06a788a2af43e.
5. Michael Woods, "Alfred W. Harris (1853–1920)," in *Encyclopedia Virginia* (Virginia Humanities, December 7, 2020), https://encyclopediavirginia.org/entries/harris-alfred-w-1853-1920/; Dr. Martin Luther King Jr. Memorial Commission, "African American Legislators in Virginia," Virginia General Assembly, accessed October 2, 2025, https://mlkcommission.dls.virginia.gov/lincoln/african_americans.html.
6. Martin Luther King Jr., *Strength to Love* (Harper & Ball, 1963; Fortress Press, 2010), 26. Citation refers to the Fortress Press edition.

Chapter 9. Fruit

1. Marc Kovac, "Jesse Jackson Uses Sports Analogy for Election," *The Daily Jeffersonian* (Cambridge, Ohio), November 3, 2016, https://www.daily-jeff.com/story/news/2016/11/03/jesse-jackson-uses-sports-analogy/18929753007.

SELECTED BIBLIOGRAPHY

Anderson, Claud. *Black Labor, White Wealth: The Search for Power and Economic Justice*. PowerNomics Corporation of America, 1994.

Anderson, Claud. *PowerNomics: The National Plan to Empower Black America*. PowerNomics Corporation of America, 2001.

Buhner, Stephen Harrod. *The Lost Language of Plants: The Ecological Importance of Plant Medicine to Life on Earth*. Chelsea Green, 2002.

Carney, Judith Ann and Richard Nicholas Rosomoff. *In the Shadow of Slavery: Africa's Botanical Legacy in the Atlantic World*. University of California Press, 2009.

Daniel, Pete. *Dispossession: Discrimination Against African American Farmers in the Age of Civil Rights*. University of North Carolina Press, 2013.

Daniel, Pete. *The Shadow of Slavery: Peonage in the South, 1901-1969*. University of Illinois Press, 1972.

Du Bois, W. E. B. *The Souls of Black Folk*. A.C. McClurg & Co, 1904.

Equal Justice Initiative. "Lynching in America: Confronting the Legacy of Racial Terror." 2017. Accessed September 30, 2025. https://lynchinginamerica.eji.org/report.

French, Howard W. *Born in Blackness: Africa, Africans, and the Making of the Modern World, 1471 to the Second World War*. Liveright, 2021.

Hacker, J. David. "From '20. and Odd' To 10 Million: The Growth of The Slave Population in the United States." *Slavery & Abolition* 41, no. 4 (2020): 840–55. https://doi.org/10.1080/0144039X.2020.1755502.

Haley, Alex. *Roots: The Saga of an American Family*. Doubleday, 1976.

Hughes, Langston. *The Collected Works of Langston Hughes*. 16 vols. University of Missouri Press, 2001–03.

Jones, Bruce A. and Edwin J. Nichols. *Cultural Competence in America's Schools: Leadership, Engagement and Understanding*. Information Age Publishing, 2013.

Lewis, Edna. *The Taste of Country Cooking*. Knopf, 1976.

Lewis, Jori. *Slaves for Peanuts: A Story of Conquest, Liberation, and a Crop That Changed History*. New Press, 2022.

Muhammad, Andrew, Christopher Sichko, and Tore C. Olsson. "African Americans and Federal Land Policy: Exploring the Homestead Acts of 1862 and 1866." *Applied Economic Perspectives and Policy* 46, no. 1 (2024) : 95–110. https://doi.org/10.1002/aepp.13401.

National Research Council. *Lost Crops of Africa*. 3 vols. The National Academies Press, 1996–2008.

Penniman, Leah. *Farming While Black: Soul Fire Farm's Practical Guide to Liberation on the Land*. Chelsea Green Publishing, 2018.

Perry, Imani. *Black in Blues: How a Color Tells the Story of My People*. Ecco, 2025.

Reynolds, Bruce J. *Black Farmers in America, 1865-2000: The Pursuit of Independent Farming and the Role of Cooperatives*. RBS Research Report No. 194. US Department of Agriculture Rural Business Cooperative Service, 2002. https://www.rd.usda.gov/files/RR194.pdf.

Rochester, Shawn D. *The Black Tax: The Cost of Being Black in America*. Good Steward Publishing, 2017.

Smith, Denise A. "Nourishing the Nation While Starving: The Underfunding of Black Land-Grant Colleges and Universities." The Century Foundation, July 24, 2023. Accessed September 25, 2025. https://tcf.org/content/report/nourishing-the-nation-while-starving-the-underfunding-of-black-land-grant-colleges-and-universities.

Twitty, Michael W. *Rice: A Savor the South Cookbook*. University of North Carolina Press, 2021.

Walters, Charles. *Minerals for the Genetic Code: An Exposition & Analysis of the Dr. Olree Standard Genetic Periodic Chart & the Physical, Chemical & Biological Connection*. Acres USA, 2024.

Washington, Booker T. *Up From Slavery: An Autobiography*. A.L. Burt Company, 1901.

Whatley, Booker T. *How To Make $100,000 Farming 25 Acres*. Rodale, 1987.

Whewell, William. *The Philosophy of the Inductive Sciences: Founded Upon Their History*. 2 vols. Cambridge University Press, 2014.

Wright, Tyrene. *Booker T. Washington and Africa: The Making of a Pan Africanist*. Global Africa Press, 2015.

Zinn, Howard. *A People's History of the United States: 1492–Present*. HarperCollins, 2003.

INDEX

Page numbers in *italic* refer to illustrations. Pages in the color insert begin with *C*.

C

D

ABOUT THE AUTHOR

Kori Price

Michael Carter Jr. is an eleventh-generation farmer in the United States and the fifth generation to steward Carter Farms, a century-old family farm in Orange County, Virginia. With a passion for preserving cultural heritage and vocational and personal excellence through agriculture, Michael conducts workshops on growing and marketing ethnic vegetables, empowering farmers and communities to embrace sustainable and culturally relevant practices.

As the founder of Africulture, a nonprofit organization dedicated to educating and expounding upon the principles, practices, plants, and people of African descent that have contributed to agriculture, Michael's work extends beyond farming to the advancement of economic equity and parity in the agricultural sector and every other facet of life. His leadership in this field has earned him recognition and respect across multiple platforms, institutions, and organizations.

Having worked in Ghana, Kenya, and Israel as an agronomist and organic agricultural consultant, Michael has developed a global perspective on sustainable agriculture and the importance of cultural understanding in agriculture. His work in these regions has focused on introducing organic farming techniques, building market networks, and supporting the transition of local farms to more sustainable models.

A natural educator, Michael also serves as a cliometrician, curriculum developer, and program coordinator for his educational platforms Hen Asem (Our Story) and Africulture. His Africulture course, presently taught at the University of Virginia, delves into the contributions of Africans and African Americans to global agriculture and the policies that have impeded progress, emphasizing the importance of incorporating practical, solution-based ideals to solve problems in agriculture and throughout the world.